The Art and Science of Lightning Protection

A lightning strike to an unprotected object or system can be disastrous – in the United States lightning is responsible for over 30 percent of all electric power failures; causes property damage resulting in insurance claims of billions of dollars; and accounts for an average of 85 fatalities a year, and probably ten times as many injuries. This accessible book describes all aspects of lightning protection at a moderately technical level and includes many illustrative drawings and photographs.

The physical behavior of lightning, various types of lightning damage, and general principles of protection are introduced. Subsequent chapters then consider specific protection of building structures; electrical and electronic equipment, power and communication lines; and objects such as humans, animals, aircraft, launch vehicles, boats, and trees. Salient aspects of the 2004 US lightning protection standard NFPA 780 and the 2006 International IEC lightning protection standards are discussed, as are non-standard and unapproved methods of lightning protection. The role of lightning detection and warning in effective protection is highlighted, and options for deflecting or eliminating lightning are considered.

This book will be essential reading for everyone involved in the business of lightning protection, including meteorologists, atmospheric scientists, architects, engineers, and fire-safety experts. It will also be of significant value to insurance practitioners and physicians.

Martin A. Uman received his Ph.D. in Electrical Engineering from Princeton University in 1961. Following positions at the University of Arizona and the Westinghouse Research Laboratories in Pittsburgh, he was appointed as a Professor at the University of Florida in 1971. He was Chair of the Department of Electrical and Computer Engineering from 1991 to 2003 and currently holds the rank of Distinguished Professor. He is generally acknowledged to be one of the world's leading authorities on lightning and is best known for his work in lightning modeling, the most notable practical spinoffs of which are the LLP lightning locating system and the redefinition of several important lightning characteristics relative to hazard protection. Professor Uman has written several previous books on lightning, including *Lightning: Physics and Effects*, co-authored by Vladimir A. Rakov (Cambridge University Press, 2003). He has also published over 200 research papers and holds four patents in the area of lightning detection and location. Professor Uman has been the recipient of numerous awards including the 2001 American Geophysical Union John Adam Fleming Medal for "*outstanding contribution to the description and understanding of electricity and magnetism of the earth and its atmosphere*" and the 1996 Heinrich Hertz Medal by the Institute of Electrical and Electronics Engineers (IEEE) for "*... outstanding contributions to the understanding of lightning electromagnetics and its application to lightning detection and protection.*"

Fig. 314. — Le paratonnerre portatif, ou le parapluie-paratonnerre de Barbeu-Dubourg.

A personal lightning protection system proposed to Benjamin Franklin by Jacques Barbue-Dubourg in a letter dated 1773. See Section 7.4 for more details.

The Art and
Science of Lightning Protection

Martin A. Uman
University of Florida

CAMBRIDGE
UNIVERSITY PRESS

CAMBRIDGE UNIVERSITY PRESS
Cambridge, New York, Melbourne, Madrid, Cape Town, Singapore,
São Paulo, Delhi, Dubai, Tokyo, Mexico City

Cambridge University Press
The Edinburgh Building, Cambridge CB2 8RU, UK

Published in the United States of America by Cambridge University Press, New York

www.cambridge.org
Information on this title: www.cambridge.org/9780521158251

First published 2008
First paperback edition 2010

A catalogue record for this publication is available from the British Library

ISBN 978-0-521-87811-1 Hardback
ISBN 978-0-521-15825-1 Paperback

To my wife Dorit,
our grown offspring Mara, Jon, and Derek (in birth order),
and our way-above-average grandchildren
Sara, Hunter, Hayden, Summer, Isabella, and Mikayla
(in birth order, age 11 to age 5)

Contents

Preface

One of the questions I've most been asked is "Do lightning rods really work?" The question can be phrased more accurately as "Does the standard lightning protection eliminate the possibility of lightning damage to structures?"; and the answer is "almost always." Nevertheless, there are many individuals who erroneously believe that installing lightning protection on a structure significantly increases the risk of that structure's being struck and hence damaged by lightning, a relatively common view during the nineteenth century when little was known about the physics of lightning. Among those individuals who still so believe, as brought to my attention recently by an investigative newspaper reporter, are important officials in at least several high-profile fire departments, individuals who should know better. There are both international and national standards for installing lightning protection systems, and in this book I examine the assumptions underlying the common approach taken in all the standards and the resultant adequacy of those standards.

I am also often asked about the efficacy of the two commercially available methods of "non-standard" lightning protection, non-standard in that they are not recommended in either the primary international standard or the US standard. Each method involves the use of unusual lightning rods (air terminals), in one case purportedly to attract the lightning from relatively far away so that fewer air terminals are needed and in the other case purportedly either to discharge the thundercloud (the latter possibility being originally proposed by Benjamin Franklin based on his laboratory experiments) or, at minimum, to emit enough electrical charge into the atmosphere surrounding the air terminal to repel lightning from the local vicinity. As you will read in Chapters 4 and 13, the evidence is that these two non-standard approaches are no better than standard protection and in some respects may well be worse.

This book is intended to describe all aspects of lightning protection at a moderately technical level in an easy-to-read format, including many illustrative drawings and photographs. It should be relatively easy reading for those with training in physics or electrical engineering. Some or all of the material in this book should also be of value to meteorologists, architects, building-construction engineers, fire-safety experts, and insurance practitioners; and I hope it will make interesting reading for high-school and college science students and other technically oriented individuals. Additionally, boaters, pilots, passengers in ships and airplanes, and outdoor recreationists will find their questions about the lightning hazard explored.

I have authored or co-authored four books on lightning, those being published in 1969, 1971, 1989, and 2003. These four books are primarily concerned with the physics of lightning and, as an indication of the general interest in lightning, are all still in print in paperback editions. I have long been interested in writing a book that contains both the known information on lightning protection and my considered opinions on that information as accumulated in 40-plus years of lightning research. *The Art and Science of Lightning Protection* draws heavily on the 2003 technical monograph *Lightning: Physics and Effects* (Cambridge University Press) that Professor V. A. Rakov and I co-authored. *The Art and Science of Lightning Protection* reorganizes and expands upon parts of Chapters 10, 18, and 19 of that book and also contains considerable new material. *Lightning: Physics and Effects* provides the best reference to the various aspects of the physics of lightning that are necessarily considered only briefly here.

This book is divided into 14 chapters. Chapter 1 provides an introduction to the physical behavior of lightning, the general principles of protection, and the statistics of lightning occurrence. Chapter 2 discusses the various types of lightning damage, and the lightning properties that produce that damage. Chapters 3, 4, and 5 are primarily concerned with aspects of the protection of structures: Chapter 3, the theoretical foundations for protection techniques; Chapter 4, the nuts and bolts of air terminals and down conductors; and Chapter 5, grounding. Chapter 6 surveys the protection of electrical and electronic equipment located inside structures. Chapters 7, 9, 10, 11, and 12 are each concerned with the protection of a specific object or system: humans and animals, aircraft and launch vehicles, boats, trees, and power and communication lines, respectively. Chapter 8 discusses lightning detection and warning, a very important factor in the protection of humans, electrical power distribution and transmission systems, and forests. Preventative action can often be taken if it is known that lightning activity is approaching or is occurring in the vicinity. Chapter 13 considers whether lightning can be eliminated or can be deflected from its intended trajectory. Finally, a summary of the status of our understanding of lightning protection is given in Chapter 14.

The manuscript was typed and retyped (word processed and re-word processed) by Waleta Newman and Kathryn Thomson. They were an invaluable resource who also handled the myriad details involved in assembling the manuscript, for which I cannot express enough appreciation. Modification of existing figures and most of the original drawings in this book were provided by University of Florida electrical engineering student Britt Hanley. Photographic and other credits for a variety of contributors are found in the figure captions. A number of chapters were graciously criticized by experts in specialized aspects of lightning. For excellent technical advice, I express my appreciation to Christopher J. Andrews, M.D., Ph.D., of Brisbane, Australia, Professor Mary Ann Cooper, M.D., of the University of Illinois at Chicago, Professor Emeritus Mat Darveniza of the University of Queensland, Australia, Professor E. Philip Krider of the University of Arizona, and Charles Williams of S&C Electric Company. I would also like to thank electrical engineering students Dustin Hill and Britt Hanley for thoroughly reading the manuscript and making many helpful suggestions. Finally, I would like to say that I am grateful to the management of the Department of Electrical and Computer Engineering at the University of Florida for providing an environment conducive to my research and to the writing of this book.

1   What is lightning?

1.1   Types of lightning

Lightning is a very long electrical spark, "very long" meaning greater than about 1 kilometer. Most lightning is generated in summer thunderstorms and is characterized by a length of 5 to 10 km, at the extreme about 100 km. The longest electrical sparks that can commonly be generated in the laboratory measure 1 to 3 meters, with the maximum being 10 to 20 m. In swirling desert sand storms, sparks occur that are a few meters in length (Kamra 1972a,b). Sparks this short are generally not called lightning. Some volcanoes produce kilometer-length electrical discharges in their ejected material, so-called "volcano lightning" (e.g., Anderson *et al.* 1965, Brook *et al.* 1974a,b, McNutt and Davis 2000). Even longer discharges, called "nuclear lightning," were produced by near-surface thermonuclear (H-bomb) explosions in the 1950s (testing that has since been discontinued), the electrical charge source for the nuclear lightning being the negative charge (electrons) blasted upward into the atmosphere by the detonation (e.g., Uman *et al.* 1972, Williams *et al.* 1988). Video records taken both by the Galileo orbiter as it circled the planet Jupiter and by earlier fly-by spacecraft document that lightning occurs in the clouds of that planet (e.g., Borucki and Magalhaes 1992, Little *et al.* 1999), and there is evidence that lightning also occurs on Saturn. Two varieties of Earth lightning that occur between cloud and ground are shown in the photographs of Figs. 1.1 and 1.2. Figure 1.1 illustrates the most common cloud-to-ground lightning, lightning that is initiated in the cloud and travels to Earth. Figure 1.2 shows a photograph of a less common variety of lightning between cloud and ground, lightning that is initiated at an object on the Earth's surface and then propagates upward toward or into the cloud charge. In both cases, the direction of branching, downward or upward, indicates the direction of propagation of the initiating discharge process.

Lightning is a distant cousin of the short electrical spark that can be created between your finger and a metal doorknob if you walk across certain rugs in a dry, cool environment; or that can be produced between your finger and your car door after you slide across a car seat on a winter day. The process of electrically charging one's body by scuffing shoes on a rug or by rubbing pants on a car seat results in a body voltage of roughly 10 000 volts (10 kV), a level of voltage that can drive an electrical spark through air a distance of about one-third of a centimeter (roughly one-eighth of an inch) between a 10 kV finger and any

Fig. 1.1   Cloud-to-ground lightning over the Arizona desert. Courtesy of J. Rodney Hastings.

uncharged conducting object, including friends. (The breakdown voltage between plane parallel electrodes separated by 1 cm at standard temperature and pressure is 30 kV.) Frictional charging (the triboelectric effect) involving rugs and shoes or car seats and pants shares common features with the charging process involving the interaction between various forms of ice and water that takes place inside thunderclouds. In the short-spark case, the rubbing of two dissimilar materials, say rubber (your shoe sole) and nylon (your rug), causes electrons to transfer from the nylon to the rubber, charging the rubber negatively (with an excess of electrons) and leaving the nylon positively charged (with a deficiency of electrons). In a thunderstorm, the charge transfer process is thought to involve collisions between (1) soft hail particles that are heavy enough to fall or remain stationary in the thunderstorm's updrafts and (2) small crystals of ice that are light enough to be carried upward in those updrafts. To produce the primary thundercloud charges that have been observed, these ice–hail interactions must take place at altitudes where the temperature is considerably colder than freezing, generally −10 °C to −20 °C, in the presence of unfrozen (super-cooled) water droplets. (The freezing temperature is 0 °C or 32 °F.) After charge has been transferred between the colliding ice and hail particles, the positively charged ice crystals are carried further upward in updrafts to the top part of the thunder-cloud, to an altitude near 10 km above sea level in temperate summer storms; while the negatively charged hail resides at an altitude of 6 to 8 km. Thus, the main charge structure of an isolated, mature thundercloud consists of many tens of coulombs of positive charge in its upper portions and a more or less equal

Fig. 1.2 Ground-to-cloud lightning. Four upward lightning flashes initiated concurrently, by visual observation, from four 300-m-tall television transmission towers during a frontal thunderstorm in Kansas City. The TV towers are located along a line 10 km long. Courtesy of C. Gill Kitterman (Kitterman 1981).

negative charge in its lower levels. (One coulomb is the charge that passes through a current-carrying wire when 1 ampere of electrical current flows for 1 second – a typical household circuit is designed to carry 15 amperes of electrical current continuously.) In a typical thundercloud, a small positive charge is also found below the main negative charge, at altitudes where the temperature is near or warmer than freezing. There are a variety of mechanisms that have been suggested to produce this lower positive charge including collisions between different types of precipitation.

The thundercloud charge structure discussed above is illustrated in Fig. 1.3, along with the potential locations of some different types of lightning flashes. Note that while the two main charge centers are labeled, the small positive charge

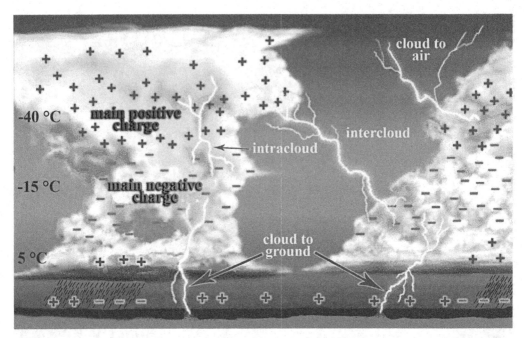

Fig. 1.3   The charge structure of two simple isolated thunderclouds and some of the locations where
lightning can occur. Adapted from *Encyclopedia Britannica Online*, Lightning (2007) (see
http://search.eb.com/eb/article-9048228).

region residing below the main negative charge is not labeled, for lack of space.
Note also that, again for lack of space, upward lightning (Fig. 1.2) is not illustrated
in Fig. 1.3; nor is cloud-to-ground lightning from either of the two positive charge
regions, nor cloud-to-air discharges from other charge regions than the main
positive; nor intracloud lightning between the main negative charge center and
the small positive charge below it. Another drawing of the thundercloud charge
distribution is given in Fig. 9.3, as part of a figure illustrating the altitude and
ambient temperature at which aircraft are struck by lightning. The charge structure
in a thunderstorm is actually more complex than shown in Fig. 1.3 or Fig. 9.3,
varies from storm to storm, and is occasionally very much different from the
structure illustrated, even upside-down with the main positive charge on the
bottom and the main negative charge on top. Further, the two isolated thunderclouds illustrated in Fig. 1.3 may form a portion of many contiguous and interact-
ing storm "cells" that make up larger storm systems. To read more about cloud
charge and cloud charging, the reader is referred to the book by MacGorman and
Rust (1998) and the references found therein.

   All lightning discharges can be divided into two categories: those that bridge the
gap between the cloud charge and the Earth, and those that do not. The latter group
as a whole is referred to as "cloud discharges" and accounts for the majority of all
lightning discharges. As illustrated in Fig. 1.3, cloud discharges that occur totally
within a single cloud (or "cell") are called intracloud lightning (thought to be the most

common cloud lightning and the most common of all the forms of lightning); those that occur between clouds are called intercloud lightning (less common than intracloud lightning); and those that occur between one of the cloud charge regions and the surrounding air are called cloud-to-air lightning. In terms of lightning protection, cloud lightning is of interest because of its interaction with airborne vehicles such as airplanes, blimps, and launch vehicles (see Chapter 9). Since most structures and animals (humans included) that are exposed to lightning and that potentially require lightning protection are found on the Earth's surface, the properties of cloud-to-ground flashes are of primary interest in designing lightning protection.

The terms "lightning flash," "lightning discharge," and "lightning" are used interchangeably in the literature, and in this book, to describe either cloud lightning or cloud-to-ground lightning. There are four types of lightning flashes that occur between the cloud and ground. The four types, illustrated separately in Fig. 1.4a,b,c and d, are distinguished from each other by the sign of the electrical charge carried in the initiation process and by the direction of propagation of the initiation process. Figures 1.4a and c show flashes referred to as downward lightning; Figures 1.4b and d depict upward lightning. About 90 percent of cloud-to-ground lightning flashes are initiated by a negatively charged, downward-propagating "leader" as shown in Fig. 1.4a. These result in the lowering of negative charge from the main negative charge region in the middle part of the cloud to the ground. A photograph of the lightning depicted in Fig. 1.4a is shown in Fig. 1.1, is illustrated in Fig. 1.3, and is discussed further in Section 1.3. About 10 percent of cloud-to-ground lightning flashes are initiated by a positively charged, downward-propagating leader as shown in Fig. 1.4c, and result in the lowering of positive charge from the cloud to the ground, either from the upper or lower positive charge regions. The remaining two types of cloud-to-ground (actually ground-to-cloud) lightning discharges (Fig. 1.4b,d) are relatively uncommon and are upward initiated from mountain-tops, tall man-made towers, or other tall objects, toward and often into one of the cloud charge regions. A photograph of such upward lightning initiated from tall towers is found in Fig. 1.2. Note that the branching shown in Fig. 1.4b,d is upward, in the direction of propagation of the initiating discharge, as in Fig. 1.2, whereas in the downward lightning of Fig. 1.4a,c the branching is downward, again in the direction of the initiating leader that propagates from the cloud charge to the Earth, as in Fig. 1.1.

1.2   Statistics on lightning occurrence

From ground-based and satellite measurements, it has been estimated that every second there are between 30 and 100 lightning flashes (both cloud and cloud-to-ground discharges) occurring around the world. At any time, worldwide, there are about 2000 active thunderstorms. In the course of one day, there are up to 9 million worldwide flashes. Clearly, the atmosphere of the planet Earth is very active electrically. The downward view from low Earth orbit above an active thunderstorm

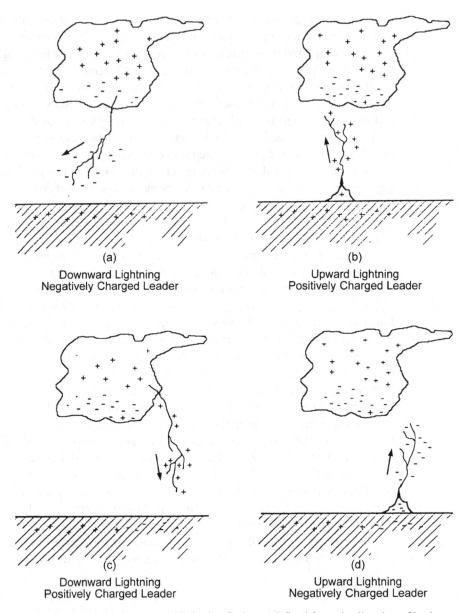

Fig. 1.4   The four types of cloud-to-ground lightning flashes as defined from the direction of leader propagation and the charge on the initiating leader. Adapted from Berger (1978).

system resembles a random sequence of flash bulbs exploding every second or less, the light from the brighter flashes illuminating their parent clouds outward radially. Video of the planet Jupiter taken from the Galileo spacecraft orbiter shows similar lightning-produced cloud illuminations although not at such a high rate as those on

Earth. On Earth, most of the lightning is over the land since heating of the land by the Sun is the most common source of the rising, heated humid air needed to initiate the thunderstorm charge generation and separation processes. In general, the hotter and more humid the local atmosphere, the more thunderstorms and the more lightning. Additionally, the air temperature must decrease with altitude in such a way as to allow the hot humid air to rise efficiently. Parts of tropical Africa and Indonesia have lightning more than 200 days per year. The Oregon and Washington coasts of the United States have virtually no lightning because the Pacific Ocean keeps the air cool, suppressing upward air motion. The central and southwest parts of Florida, which in most years experience the highest number of lightning flashes per square kilometer in the United States, have lightning about 90 days per year.

If we assume that there are 100 combined cloud and cloud-to-ground flashes worldwide per second, a reasonable upper limit to the various estimates, the average "flash density" (the number of cloud and cloud-to-ground flashes in or over a given area of the Earth each year) is about 6 per square kilometer per year ($km^{-2} yr^{-1}$) or about 16 per square mile per year. (There are about 2.6 square kilometers in a square mile.) Since most of the flashes are over land and most of the Earth's surface is covered with water, the actual over-land average annual flash density is considerably higher than $6 km^{-2} yr^{-1}$. Along the southwest coast of Florida the "ground flash density" (the number of cloud-to-ground flashes striking a given area of the Earth each year) is near $15 km^{-2} yr^{-1}$, or about 40 per square mile per year. Since lightning within half a kilometer or so produces very loud thunder that arrives within a second or so of the light (see Section 8.1), people living along Florida's southwest coast should expect to hear about 10 to 20 house-rattling thunders each year, almost all in the summer time. In fact, according to the statistical theory of Krider (2005), in any given year, for a ground flash density of $15 km^{-2} yr^{-1}$ there is a 90 percent chance that at least one random ground flash will be within 220 m of a fixed location (there will probably be more), a 50 percent chance that at least one random flash will be closer than 120 m, and a 10 percent chance that at least one random flash will be within about 50 m. A map of the ground flash density for the continental United States, compiled from a network of ground-based radio frequency sensors (see Chapter 8), is found in Fig. 1.5. A world flash density map derived from satellite data is given in Fig. 8.8. According to the data from which Fig. 8.8 was derived, the Congo basin in Africa includes an area of over $3 \times 10^6 km^2$ that exhibits total flash densities (cloud and cloud-to-ground flashes) greater than the value for the Tampa, Florida area. In Rwanda, the total flash density is about $80 km^{-2} yr^{-1}$.

A typical small thunderstorm system produces a lightning flash to ground every 20 to 30 seconds for 40 to 60 minutes and covers an area of typically 100 to 300 square kilometers (about 40 to 115 square miles), roughly a circle on the ground with a radius between 6 and 10 km, or 4 to 6 miles. Large storm systems can produce more than one flash to ground each second over areas a hundred times or more larger.

A typical house in Florida will be struck by lightning about once every 50 years. Said another way, one out of every 50 or so houses in Florida will be struck each

Fig. 1.5   A map of the average cloud-to-ground flash density (in flashes per square kilometer per year) in the continental United States from 1989 through 1998. Courtesy of Vaisala, Inc.

year. Often there is little or no damage; sometimes there is total destruction by fire. A rough calculation to support this expected strike rate to Florida residences is found in Section 1.5.

The Empire State Building in New York City, a structure whose height is about 300 m (roughly 1000 feet), gets struck by lightning 20 to 25 times per year, with about 80 percent of this lightning being of the upward type (Fig. 1.4b,d; Fig. 1.2). Towers of similar height in Florida get struck about 100 times per year. Towers of height less than 100 m standing on flat ground are struck mostly by downward lightning (Fig. 1.4a,c; Fig. 1.1), while towers over 400 to 500 m high are struck mostly by upward lightning. Since such tall towers initiate the upward lightning, the word "struck" may not be appropriate. In general, a tower that is twice as tall as another one in the same general location will be struck about four times as often; and for a given tower height, if the ground flash density is twice as high, the strike rate will be roughly doubled.

1.3   Cloud-to-ground lightning

The most common cloud-to-ground discharge, downward lightning carrying negative charge, may well begin as a local discharge between the bottom of the main negative charge region and the small positive charge region beneath it (see Fig. 1.3). This local discharge would serve to provide free mobile electrons, those electrons being previously immobilized by virtue of their attachment to hail and other heavy particles. (Electrons carry the smallest known unit of negative electrical charge, 1.6×10^{-19} coulombs, and have less than a thousandth of the mass of the smallest atom.) Because of the electron's small mass, free electrons are extremely mobile (they move easily when exposed to an electric force) compared with the heavier air atoms or molecules that are missing an electron (have become "ionized"), or with charged hail, ice, or water particles which are essentially stationary on the timescale of lightning. Hence free electrons are the primary contributor to the lightning current. In negative cloud-to-ground lightning, the free electrons over-run the lower positive charge region, neutralizing most of its small positive charge, and then continue their trip toward ground. The physical mechanism for moving the negative charge to Earth is an electrical discharge called the "stepped leader." This process, and other salient aspects of the negative cloud-to-ground flash, are illustrated in Fig. 1.6, Fig. 1.7, and Fig. 1.8. The pioneering work in identifying the main features of cloud-to-ground lightning took place in South Africa starting in the 1930s (e.g., Schonland 1956, Malan 1963).

The stepped leader's movement from cloud-to-ground is not continuous. Rather, it moves downward in discrete luminous segments of about 50 m length, then pauses, then moves another 50 m or so, and so on. Each added length the leader forges is called a step. In Fig. 1.6 the luminous steps appear as darkened tips on the less-luminous leader channel extending downward from the cloud. Each luminous leader step appears in a microsecond (a millionth of a second or μs) or less. The time between luminous steps is about 50 μs when the stepped leader is far above the ground and less, near 10 μs, when it is close to the ground. The downward-propagating stepped leader branches downward, as noted earlier. Negative charge in the form of electrons is more-or-less continuously lowered from the main negative charge region in the middle of the cloud (Fig. 1.3) into the leader channel. The average downward speed of the bottom of the stepped leader during its trip toward ground is about $2 \times 10^5 \, \text{m s}^{-1}$ (200 kilometers per second) with the result that the trip between the cloud charge and ground takes about 20 milliseconds (thousandths of a second or ms). A typical stepped leader has about 5 coulombs of negative charge distributed over its length when it is near ground. To establish this charge on the leader channel an average current of about 100 to 200 amperes must flow during the whole leader process. The pulsed currents which flow in generating the leader steps have a peak current of the order of 1000 amperes. Each leader step produces a pulse of visible light, a pulse of radio frequency energy, and a pulse of X-rays (Dwyer *et al.* 2005). The luminous diameter of the stepped leader has been measured photographically to be between 1 and 10 m. It is thought, however, that most of the stepped-leader current flows down a narrow conducting core a few centimeters in diameter at the center of the observed leader. The large photographed diameter is probably due to a luminous "corona," a low-level electrical discharge surrounding the conducting core.

When the stepped leader is near the ground, its relatively large negative charge induces (attracts) concentrated positive charge on the conducting Earth beneath it and especially on objects projecting above the Earth's surface. If the attraction

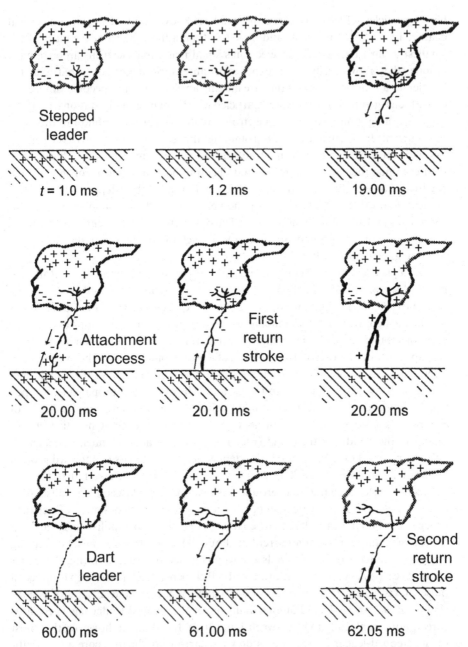

Stepped
leader

Fig. 1.6   A drawing depicting the development of a negative cloud-to-ground lightning flash, the most common type of cloud-to-ground lightning. The timescale is given in milliseconds (ms) from the first electrical breakdown processes in the cloud. Adapted from Uman (1987, revised 2001).

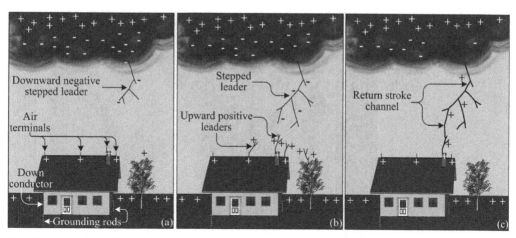

Fig. 1.7 A closer view of the attachment process and the successful capture of the flash by a lightning rod. The time from (a), showing the final stages of the downward-moving, negatively charged stepped leader, to (b), showing upward-moving, positively charged leaders initiated from two lightning rods and the tree, is perhaps 0.5 ms, and from (b) to (c), in which the return stroke has occurred, is about 0.2 ms.

between the opposite charges is strong enough, the positive charge on the Earth or Earth-bound objects will attempt to join and neutralize the negative charge above. The mechanism for doing so is the initiation of upward-going electrical discharges from the ground or from grounded objects, as illustrated in Fig. 1.6 at 20.00 milliseconds, in Fig. 1.7 (specifically 1.7b), and in Fig. 1.8, and as discussed further in Section 3.3. One of these upward-going discharges contacts a branch of the downward-moving leader and thereby determines the lightning strike-point and the primary lightning current path (channel) between cloud and ground. Figure 1.8 shows perhaps the best scientific measurement of this "attachment process" along with the resulting return stroke current. The time-resolved "streak" photograph was obtained by Berger and Vogelsanger (1966) using a camera in which film is moved continuously and horizontally behind a stationary lens with the camera shutter open. Thus the image of each luminous step of the stepped leader is displaced horizontally on the film from the previous step. The final 300 μs or so of the stepped leader is evident in Fig. 1.8 starting in the upper left corner and progressing downward to point A, the termination of the downward stepped leader. An upward-connecting leader (not visible on the photograph, presumably because of the relatively low luminosity of positive discharges), which is probably stepped, evidently rises from the top of the 55 m tower and splits at point B, one branch going upward to the left and the other branch going upward to the right and connecting to a downward-moving discharge from the end of the stepped leader at point A. That final connection, illuminated by the following "return stroke" (see next paragraph) takes place somewhere between A and B, perhaps close to A judging from the length of the unconnected upward left branch. The last observed downward leader step is slightly over 30 m above the tower top and a little over 30 m

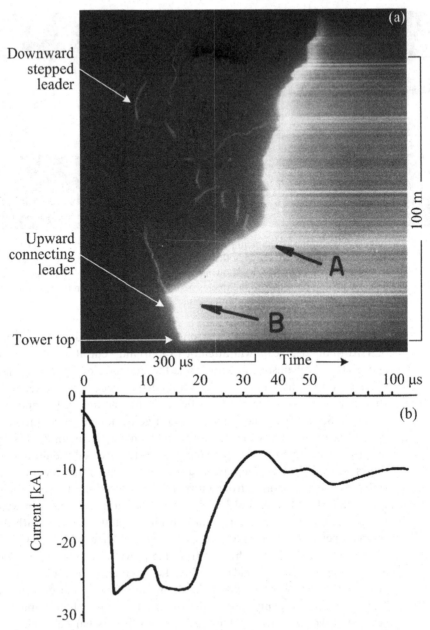

Fig. 1.8   Streak camera photograph of the attachment process of a downward lightning flash carrying negative charge to a 55 m tower on top of Mount San Salvatore in Lugano, Switzerland. The images of downward-stepped leader steps visible to point A are enhanced for clarity. Shown below the streak image is the return stroke current waveform (on a logarithmic scale) measured on the top of the tower, peak current being 27 kA. Adapted from Berger and Vogelsanger (1966).

away horizontally. The straight line distance from the last leader step at point A to the tower top is 37 m, but the attachment process chooses a longer, more circuitous path. While we are assuming above that the attachment process takes place below the last photographed leader step, point A, it is possible that some of the final leader steps could also be part of the attachment process. Understanding the attachment process is critical to the proper design of lightning protection systems. Unfortunately, the attachment process is sufficiently complex and variable that it has not been possible to gain more than a relatively crude understanding of it.

When a downward-moving negatively charged leader branch and an upward-moving positively charged leader connect, negative charge near the bottom of the leader channel moves violently downward into the Earth, causing large currents to flow at ground and causing the channel near ground to become very luminous. Since electrical signals (or any signals, for that matter) have a maximum speed of 3×10^8 m s^{-1} (300 000 kilometers per second or 186 000 miles per second) – the "speed of light" – the leader channel above ground has no way of knowing for a short time that the leader bottom has been connected to ground and has become highly luminous and highly electrically conducting. The channel luminosity and current, in a process termed the first return stroke, propagate continuously up the channel and out (down) the branches of the leader channel at a speed typically between one-third and one-half the speed of light, as illustrated in Fig. 1.6 at 20.10 and 20.20 ms, and in Fig. 1.7c. In Fig. 1.8a the return stroke luminosity illuminates the 100 m portion of the path of the stepped leader and attachment process shown, in a turn-on time of about 1 μs (starting around 150 μs), and thereafter the channel luminosity is streaked to the right. Even though the return stroke's high current and high luminosity move upward on the main channel, electrons at all points in the main channel always move downward and represent the primary components of the current. Electrons flow up the branches toward the main channel while the return stroke traverses the branches in the outward and downward direction. Eventually, some milliseconds after the return stroke is initiated, the several coulombs of negative electric charge which were resident on the stepped leader all flow into the ground. Additional current may also flow to ground directly from the cloud once the return stroke has reached the cloud.

The return stroke produces the bright channel of high-temperature air that we see. The maximum return stroke temperature is near 30 000 °C (50 000 °F). We usually do not see the dimmer, downward-moving stepped leader with our eyes but can record it with high-speed cameras (Fig. 1.8a). The reason we do not visually detect the stepped leader preceding a first return stroke is apparently because the eye cannot resolve the time between the formation of the weakly luminous leader and the explosive illumination of the leader channel by the return stroke. The human eye also cannot respond quickly enough to resolve the upward propagation of the return stroke, and thus it appears as if all points on the return stroke channel become bright simultaneously. The return stroke impulsively heats the current-carrying air which then expands and thereby produces most of the thunder we hear.

After the first stroke current has ceased to flow, the lightning flash may end, in which case the discharge is called a single-stroke flash. About 80 percent of flashes that lower negative charge to ground in temperate regions contain more than one stroke. Three to five strokes are common. The individual strokes are typically separated by 40 or 50 ms. Strokes subsequent to the first (called "subsequent strokes") are initiated only if additional negative charge is made available to the upper portion of the previous stroke channel in a time less than about 100 ms from the cessation of the current of the previous stroke. When this additional charge is available, a continuously propagating leader (as opposed to a stepped leader), known as a "dart leader," moves down the defunct return stroke channel, again depositing negative charge from the negative charge region along the channel length, as illustrated in Fig. 1.6 at 60.00 and 61.00 ms. The dart leader thus sets the stage for the second (or any subsequent) return stroke. The dart leader's earthward trip takes a few milliseconds. To high-speed cameras the dart leader appears as a luminous section of channel tens of meters in length which travels smoothly earthward at about 1/30 the speed of light (about $10^7 \mathrm{m\,s}^{-1}$). The dart leader generally deposits somewhat less charge, perhaps a tenth as much, along its path than does the stepped leader, with the result that subsequent return strokes generally lower less charge to ground and have smaller peak currents than first strokes.

Subsequent stroke peak currents are typically 10 000 to 15 000 amperes (10 to 15 kA), while first stroke currents are typically near 30 kA. A first return stroke current with peak value 27 kA is shown in Fig. 1.8b. Return stroke currents from a three-stroke flash that struck a 60 m tower in South Africa are shown in Fig. 1.9. The first stroke in that particular flash had a peak current about twice the typical, the second stroke about five times the typical peak current of a "subsequent" stroke, while the third stroke had a typical current for a stroke following the first (see Table 2.1). The rise times (typically measured between 10 percent and 90 percent of peak value as illustrated in Fig. 2.2) of subsequent stroke currents (see Fig. 2.2) are generally less than 1 μs, often tenths of a microsecond, whereas current rise times for the first stroke are some microseconds (Fig. 1.8, Fig. 1.9, Table 2.1).

The first return stroke in a negative cloud-to-ground flash appears to be strongly branched downward because the return stroke follows the path and branches of the previous stepped leader. Dart leaders generally follow only the main channel of the previous stroke and hence subsequent return strokes generally exhibit little branching. There is a leader that is intermediate between the stepped leader and the dart leader. Dart leaders propagating down the remains of more-decayed (either older or subjected to more wind turbulence) or less-well-conditioned return stroke channels may at some point begin to exhibit stepping, either within the confines of those channels of warm, low-density air or leaving those channels and propagating into virgin air, in either case becoming so-called dart-stepped leaders. Because some of the dart-stepped leaders form new paths to ground, one-third to one-half of all lightning flashes to ground contact the Earth in more than one location.

A typical cloud-to-ground discharge lowers about 30 coulombs of negative charge from the main negative charge region of the cloud (Fig. 1.3) to the Earth.

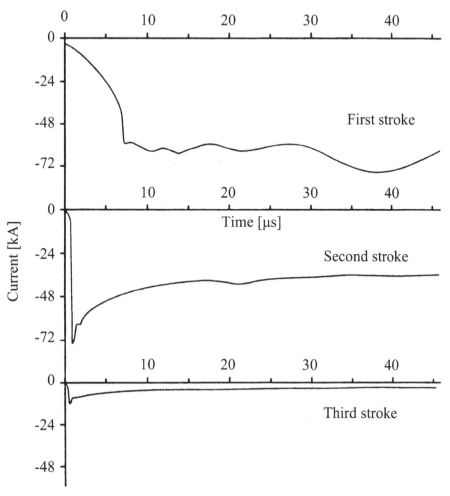

Fig. 1.9   Return stroke current observed at ground for a first stroke and two subsequent strokes lowering negative charge to a 60 m tower standing on flat ground in South Africa. The first two strokes are unusually large. Adapted from Eriksson (1978).

This charge is transferred in a few tenths of a second by the several strokes and any "continuing current" which may flow from the cloud charge source to ground after a stroke. Most continuing current follows subsequent strokes. Half of all flashes contain at least one continuing current interval exceeding about 40 ms. The time between strokes that follow the same channel can be as long as tenths of a second if a continuing current flows from the cloud charge into the channel after a given stroke. Apparently, the channel is receptive to a new dart leader only after all current, including continuing current, has terminated. While the leader/ return stroke process transfers charge to ground in two steps (charge is put on the leader channel from the cloud charge, from the top down, and then is discharged to ground from the channel bottom upward), the continuing current represents a relatively steady charge flow between the main negative charge region and ground.

Thus far in this section we have discussed the usual stepped leader which lowers negative charge from the cloud to the Earth. As noted earlier, about 10 percent of cloud-to-ground flashes are initiated by downward-moving stepped leaders that lower positive charge (Fig. 1.4c), either from the main upper positive charge region of the cloud or from the small lower positive charge region (Fig. 1.3). The steps of positive stepped leaders are less distinct than the steps of negative stepped leaders. Positive return strokes can exhibit currents at the ground whose peak value can exceed 300 kA, considerably larger than for negative strokes whose peak currents rarely exceed 100 kA. Nevertheless, typical positive peak currents are similar to typical negative peak currents, near 30 kA. Positive discharges usually exhibit only one return stroke, and that stroke is almost always followed by a relatively long period of continuing current. The overall charge transfer in positive flashes can considerably exceed that in negative flashes. Although positive flashes are less common than negative flashes, the potentially large peak current and potentially large charge transfer of the positive flashes make them a special hazard that must be taken into account when designing lightning protection. More information on positive and negative current characteristics is found in Table 2.1.

In upward lightning (Fig. 1.4b,d), the first leader propagates from ground to cloud but does not initiate an observable return stroke or return-stroke-like process when it reaches the cloud charge. Rather, the upward leader primarily provides a connection between the cloud charge region and the ground. After that connection is made and the initial current has ceased to flow, "subsequent strokes," initiated by downward-moving dart leaders from the cloud charge and having the same characteristics as strokes following the first stroke in cloud-to-ground lightning may occur. About half of all upward flashes exhibit such subsequent return strokes. Natural upward lightning is similar to the upward lightning that can be artificially initiated (triggered) using the rocket-and-wire technique, as discussed in Section 13.2. The reason that the upward first leader in both natural upward and triggered lightning does not produce a detectible downward return stroke is likely to be because there is no well-defined region of vastly different electrical potential from that of the leader in the cloud, as there is in the case of the downward leader of negative potential 10^7 to 10^8 volts striking the zero-electrical-potential Earth.

For more information regarding the physics of all four types of cloud-to-ground lightning, the reader is referred to the books by Rakov and Uman (2003) and Uman (1987, revised 2001).

1.4   General principles of protection

In the previous section we have described how the downward-moving stepped leader, as it approaches ground, causes upward-moving discharges to be initiated from objects on the ground or, absent those objects, from the ground (or water) itself. One or more of these upward discharges will connect to a branch or branches of the downward leader (as part of the attachment process) determining the path of the return stroke current that flows after the connection of the upward and downward leaders, as illustrated in Fig. 1.6, Fig. 1.7, and Fig. 1.8.

Lightning protection of a structure, a tree, a power line, or of any other object attached to the Earth, can be achieved by providing an elevated, well-grounded conductor above the object so that the lightning can preferentially strike the elevated conductor, thereby shunting the lightning current to ground away from the protected object. The most common examples of elevated protective conductors are overhead ground wires on transmission power lines (Fig. 1.10, Fig. 12.2, and Fig. 12.3) and lightning rods on structures (Fig. 1.7, Fig. 1.11, Figs. 3.3–3.5, Figs. 4.4–4.9). In both of these cases, the intended strike point, an overhead ground wire or a lightning rod, is connected by vertical wires, called "down conductors" (or "down leads") to a buried "grounding electrode" (a ground rod or a series of such rods, a buried horizontal wire, or other buried conducting structure) that allows the lightning current to flow relatively harmlessly into the Earth (see Chapter 5). In providing the preferential lightning strike point, a lightning protection system draws to itself (via the upward-connecting leader), and thereafter harmlessly disposes of, the lightning current that would otherwise probably have struck the protected object. The probability of a lightning strike to the general region of a protected object is only marginally increased, if increased at all, by virtue of the presence of the lightning protection system. Thus the not uncommon belief that the presence of a lightning protection system significantly increases the likelihood of a structure's being struck and damaged is not true. The benefits of having lightning protection far outweigh any marginal increase in the lightning strike probability.

Besides the structural protection of objects discussed above, the electrical power to and within structures and the electronic equipment located within structures, such as televisions and computers, should be protected from (1) the voltages induced in those electrical and electronic systems by lightning current flowing in the lightning protection system, and the electromagnetic effects of very close lightning either attached or not attached to the protection system, and (2) the voltages resulting from lightning-induced signals entering the structure via power lines and communication lines, given that these utility lines may have relatively large voltages induced on them by direct or nearby strikes. Protection of electronics is accomplished by shielding from electromagnetic fields, filtering out the damaging high-frequency currents and voltages due to lightning, and using surge-limiting devices (often called surge protective devices or SPDs) such as spark gaps and metal oxide varistors (MOVs), as discussed briefly in Section 3.2 and more completely in Chapter 6. Similarly, overhead and underground power and communication lines may be protected from excessive voltages, which may lead to flashover and outages, by SPDs generally called "lightning arresters," most often of the MOV type (see Chapter 12).

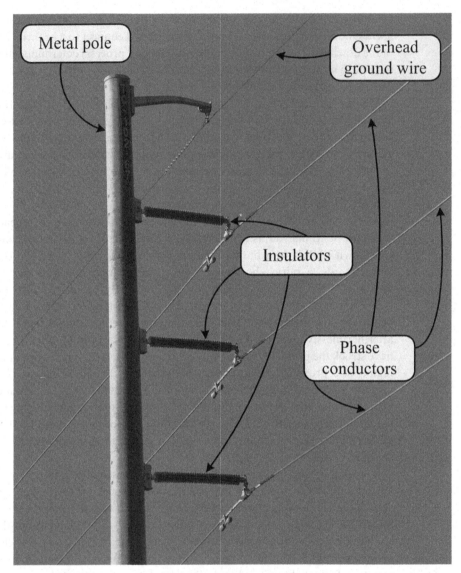

Fig. 1.10   A photograph of a 115 000 volt (115 kV) power line protected from the effects of lightning by an overhead ground wire (top wire) that is designed to intercept the lightning and conduct its current to ground via the metallic poles to which the ground wire is attached. The three vertically oriented wires beneath the overhead ground wire (the phase wires) are electrically insulated from the metallic poles and carry the high voltage. The little "dumbbells" on the phase wires are vibration dampers. Annotated photograph by Derek Uman and Jens Schoene.

Fig. 1.11 A nineteenth-century lightning rod (air terminal) located on the roof of a residence in Florida. The two glass balls and the finial, the star-like structure on top, are decorative. Photograph by Derek Uman.

1.5   Estimation of strike probability for small structures

We now present an estimate of how often a small ground-based structure will be struck by cloud-to-ground lightning. In Chapter 3 we will look in more detail at the history and physics behind this estimate.

A structure will be struck by lightning if an upward (and not necessarily vertical) leader from that object connects with a branch of the downward-propagating stepped leader. For objects of height less than about 100 m, as we have noted, essentially all lightning strikes are expected to be downward-propagating (Fig. 1.4a,c), that is, initiated by downward-moving leaders from the cloud charge. Thus, how often a structure is struck depends on the horizontal distance from the structure edge over which the attachment process can occur. For a rough calculation, let us assume that any downward leader that has randomly propagated to a location directly above a structure or within a horizontal distance from the edge of a structure equal to twice the structure height will be intercepted by an upward and outward connecting leader. This situation is illustrated in Fig. 1.12 for a square structure of horizontal side s meters and height h meters. In this case a downward leader descending over an area equal to $(s + 4h)^2$ square meters, shown shaded in Fig. 1.12 and called the "equivalent collective area" (NFPA 780:2004) or the "collection area" (IEC 62305-2:2006), will strike the

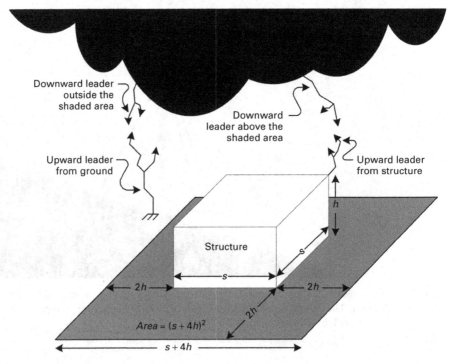

Fig. 1.12   A drawing illustrating the equivalent collective area on the ground, shown shaded, for a small structure.

structure whose actual cross-sectional area is s^2 square meters. If there are N_g flashes to ground per square kilometer per year, the number of ground flashes that strike the structure is $N_g(s+4h)^2$ where the area $(s+4h)^2$ must now be expressed in square kilometers if those units are used for N_g. In parts of northern Florida the ground lightning flash density is $10\,\text{km}^{-2}\,\text{yr}^{-1}$ (see Fig. 1.5). If the square structure is 20 m on a side and 5 m high, $(s+4h)^2 = (20+4 \times 5)^2\,\text{m}^2 = 1600\,\text{m}^2 = 0.0016\,\text{km}^2$ (1 square meter equals one-millionth of a square kilometer). Thus $N_g \times 0.0016 = 0.016 \approx 0.02$ strikes to the structure per year. Two-hundredths (one-fiftieth) of a strike per year is equivalent to one strike every 50 years, on average. Of course, one does not know when the first strike will occur. The crude calculation given above is probably only accurate to a factor of two or so; that is, the correct long-term average could be two times larger or two times smaller than the value calculated. Uncertainties in the calculation include (1) our inadequate knowledge of exactly how far horizontally from the edge of the structure the attachment process can take place, a distance that is likely to be different for each lightning event, so our assumed value should be considered some sort of average, and (2) the fact that one-third to one-half of flashes strike the ground in two or more locations.

If the ground flash density is one-fifth of the value of 10 given above for parts of northern Florida, that is, $N_g = 2$, characteristic of much of the northeastern United States (Fig. 1.5), there will be one strike to the structure every 250 years. Further, if the structure is larger than the size assumed above, it will suffer more strikes, as evidenced by the "equivalent collective area" calculation. The number of strikes per year to a given length of power line can be similarly calculated using the technique of equivalent collective area (see Section 12.1).

The justification for using a horizontal capture distance twice the structure height is, in part, experience showing that the structure strike rate calculated with this assumption is not unreasonable. Other values for the horizontal capture distance are found in the literature. In particular, the International Electrotechnical Commission, which publishes an international standard for lightning protection in five volumes (IEC 62305-1, 2, 3, 4, 5:2006), has adopted a horizontal capture distance of three times the structure height for small structures. That value has been further adopted by a number of national standards including the United States lightning protection standard (NFPA 780:2004). The various assumptions involved in determining the horizontal capture distance as a function of structure height (or the ratio, called the protective ratio, of the horizontal capture distance to the structure height) and hence the equivalent collective area of a structure are examined in Sections 3.3, 3.4, and 4.2. From a practical point of view, the exact value chosen for the capture distance is not critical to the design of adequate lightning protection.

1.6 When is lightning protection needed and/or desirable?

The cost of commercially installed lightning protection for the structure of a typical residential house is a few thousand US dollars. Often included in the overall fee is some degree of protection for electrical power and telecommunication services, these generally representing a small fraction of the cost of the structural protection. Lightning damage to structures ranges from minor to total destruction by fire. Similarly, electronics within structures may be unaffected, damaged but repairable, or destroyed by a lightning strike directly to the structure or a strike on or near the incoming service wires. Whether a personal residence should be lightning protected depends on (1) the likelihood that the structure or its power and telecommunications services will encounter lightning and (2) the personal comfort level of the residents in risking the potential deleterious effects of that lightning on the structure, to the electronics within, and even to the occupants themselves. Lightning damage to a residential structure and its contents is generally covered by homeowner's insurance, with some initial deductible, whether or not there is lightning protection on the structure. The likelihood that an isolated structure on flat ground will be struck by lightning has been considered in Section 1.5. Such a simple calculation can be extended to cover strikes on or near the service wires (see Section 12.1). The primary factors in such a calculation are the local ground flash density, and the equivalent collective area of the structure and its service wires. Other factors include the level of exposure and the environment of the structure; for example, a large isolated house on a mountain top generally presents a greater risk regarding lightning than a house in a nearby valley surrounded by tall trees.

Farm structures such as wooden barns that are particularly susceptible to lightning fires are often required to have structural lightning protection before they are insurable. As noted above, this is generally not the case for residential structures.

While national lightning protection standards such as the United States' NFPA 780:2004 generally do not have the force of law, they are often included as elements of the building construction plans for government buildings, hospitals, large industrial structures, museums, structures of cultural value, and buildings containing explosives or flammable materials. That is, lightning protection is generally mandated for structures inside which large numbers of people will congregate, for which continuation of operation is important for cultural, health, or economic reasons, or for which lightning presents a particular hazard.

It is possible to perform a detailed "risk" analysis on any particular structure in any particular location to decide in a quantitative (if somewhat arbitrary) way if lightning protection is needed (e.g., IEC 62305-2:2006; NFPA 780:2004, Annex L). The two most important and accessible factors in the analysis are the ground flash density and the equivalent collective area of the structure and its services. If the ground flash density is not available at a given location from maps like that in Fig. 1.5, it can be estimated as 0.2 times the annual number of days on which thunderstorms occur at that location, a statistic kept by weather stations worldwide. The other factors to which weight is assigned in calculating risk are generally subjective or otherwise difficult to quantify. These include the effect of the topological location of the structure within the general region (e.g., the increased risk of an isolated house on a hill top), the environment (e.g., the decreased risk of a structure surrounded by many tall trees), the value of the contents (e.g., a Rembrandt oil painting vs. a print by a local artist), the occupancy (e.g., the hazard posed by large groups of people who might, for instance, panic and trample each other), and the potential consequences to services and the environment (e.g., an increased risk provided by a structure that might explode in the middle of a major city, or by a bank or credit card company that might be shut down for a few days). After assigning a weight to each of these less easily quantified factors and knowing the probability that lightning will strike, a decision can be made over whether to provide lightning protection. The yes/no cutoff level is largely arbitrary. Risk calculations are used to justify (or "officially" show there is no need for) protection. It would seem reasonable, however, that buildings should have lightning protection if they are serving the public in areas where the ground flash density is not negligible. For personal residences, the decision of whether to have lightning protection is essentially an individual or family decision with potential personal consequences. Efforts have been and are being made to include structural and/or service lightning protection in the building codes for residential buildings, that is, to make lightning protection mandatory for residential houses. In a few municipalities and countries (e.g., Austria) this effort has been successful.

References

Anderson, R., Bjornsson, S., Blanchard, D. C. *et al.* 1965. Electricity in volcanic clouds. *Science* **148**: 1179–1189.

Berger, K. 1978. Blitzstrom-Parameter von Aufwärtsblitzen. *Bull. Schweiz. Elektrotech. Ver.* **69**: 353–360.

Berger, K. and Vogelsanger, E. 1966. Photographische Blitzuntersuchugen der Jahre 1955–1965 auf dem Monte San Salvatore. *Bull. Schweiz. Elektrotech. Ver.* **57**: 599–620.

Borucki, W. J. and Magalhaes, J. A. 1992. Analysis of Voyager 2 images of Jovian lightning. *Icarus* **96**: 1–14.

Brook, M., Moore, C. B. and Sigurgeirsson, T. 1974a. Lightning in volcanic clouds. *J. Geophys. Res.* **79**: 472–475.

Brook, M., Moore, C. B. and Sigurgeirsson, T. 1974b. Correction to "Lightning in volcanic clouds". *J. Geophys. Res.* **79**: 3102.

Dwyer, J. R., Rassoul, H. K., Al-Dayeh, M. *et al.* 2005. X-ray bursts associated with leader steps in cloud-to-ground lightning. *Geophys. Res. Lett.* **32**: L01803, doi:10.1029/2004GL021782.

Eriksson, A. J. 1978. Lightning and tall structures. *Trans. S. Afr. IEE* **69**: 238–252.

IEC 62305-1:2006. *Protection Against Lightning.* Part 1: *General Principles.* Geneva: International Electrotechnical Commission.

IEC 62305-2:2006. *Protection Against Lightning.* Part 2: *Risk Management.* Geneva: International Electrotechnical Commission.

IEC 62305-3:2006. *Protection Against Lightning.* Part 3: *Physical Damage to Structures and Life Hazard.* Geneva: International Electrotechnical Commission.

IEC 62305-4:2006. *Protection Against Lightning.* Part 4: *Electrical and Electronic Systems Within Structures.* Geneva: International Electrotechnical Commission.

IEC 62305-5:2006. *Protection Against Lightning.* Part 5: *Services* (to be published). Geneva: International Electrotechnical Commission.

Kamra, A. K. 1972a. Visual observation of electric sparks on gypsum dunes. *Nature* **240**: 143–144.

Kamra, A. K. 1972b. Measurements of electrical properties of dust storms. *J. Geophys. Res.* **77**: 5856–5869.

Kitterman, C. G. 1981. Concurrent lightning flashes on two television transmission towers. *J. Geophys. Res.* **86**: 5378–5380.

Krider, E. P. 2005. On quantifying the exposure to cloud-to-ground lightning. *International Conf. Lightning and Static Electricity* (ICLOSE 2005), Seattle, Washington, Sept. 19–23.

Little, B., Anger, C. D., Ingersoll, A. P. *et al.* 1999. Galileo images of lightning on Jupiter. *Icarus* **142**: 306–323.

MacGorman, D. R. and Rust, W. D. 1998. *The Electrical Nature of Storms.* Oxford: Oxford University Press.

McNutt, S. R. and Davis, C. M. 2000. Lightning associated with the 1992 eruptions of Crater Peak, Mount Spurr Volcano, Alaska. *J. Volcanol. Geoth. Res.* **102**: 45–65.

Malan, D. J. 1963. *Physics of Lightning.* London: The English Universities Press Ltd.

NFPA 780:2004. *Standard for the Installation of Lightning Protection Systems.* Quincy, MA: National Fire Protection Association.

Rakov, V. A. and Uman, M. A. 2003. *Lightning: Physics and Effects.* Cambridge: Cambridge University Press.

Schonland, B. F. J. 1956. The lightning discharge. In *Encyclopedia of Physics*, Vol. 22. Berlin: Springer-Verlag, pp. 576–628.

Uman, M. A. 1987. *The Lightning Discharge.* London: Academic Press. (Revised paperback edition, 2001. New York: Dover Publications, Inc.).

Uman, M. A., Seacord, D. F., Price, G. H., and Pierce, E. T. 1972. Lightning induced by thermonuclear detonations. *J. Geophys. Res.* **77**: 1591–1596.

Uman, M. A. 1986. *All About Lightning.* New York: Dover Publications, Inc. (Revised paperback edition of *Understanding Lightning*, Carnegie, PA: BEK Technical Publications, 1971).

Williams, E. R., Cooke, C. M. and Wright, K. A. 1988. The role of electric space charge in nuclear lightning. *J. Geophys. Res.* **93**: 1679–1688.

2 Lightning damage

2.1 Overview: the cost of various types of lightning damage

In the early evening of July 13, 1977, lightning struck one or more 345 000 volt (345 kV) transmission lines located about 30 miles north of New York City. Those transmission lines supplied electrical power to the New York City metropolitan area. A chain reaction of electrical events ensued with the result that electricity was lost to all of the City by 9:34 p.m. It was the next night, the night of July 14, before all electrical power was restored. About 9 million people spent a night and a day without any electricity. By the time that the blackout ended, nearly 2000 stores and businesses had been looted or damaged, and more than 3500 people had been arrested. The cost of this lightning-caused blackout has been estimated at $350 million (Sugarman 1978). A detailed technical description of its causes is found in Wilson and Zarakas (1978) and a complete bibliography of publications concerning the blackout (including testimony before Congressional committees, books, and magazine articles) is found at http://chnm.gmu.edu/search_results.php?query=new%20york%20city%20blackout.

In Minnesota on June 25, 1998, two separate lightning flashes struck and caused the de-energization of two separate 345 kV transmission lines, initiating the cascading removal of overloaded lower voltage lines from service. In a short period of time, significant portions of Minnesota, Montana, North Dakota, South Dakota, Wisconsin, Ontario, Manitoba, and Saskatchewan were without electrical power and remained so for about 19 hours. The power lost (about 1000 megawatts) was roughly one-sixth of the power of the New York City blackout discussed above but was intended to supply a much larger and less populated area.

While few power failures are as dramatic or as costly as the New York City blackout of 1977 or the mid-continental blackout of 1998, it has been estimated that about 30 percent of all power outages in the United States are lightning-related. Typically, these are more local and have a duration of minutes to hours. The total annual cost of these outages has been estimated to be near $1 billion.

Lightning causes death and injury to individuals and damage to property (e.g., Holle *et al.* 1996, Curran *et al.* 2000; see Chapter 7). Estimates of annual insurance pay-outs from lightning-related damage claims in the United States range from about one-third of a billion dollars to one billion dollars. If the former estimate is correct, given that there are typically 20 to 30 million ground flashes in the United States each year, the average insurance payment per ground flash is

$10 to $15, and there is one lightning-related insurance claim for about every 60 ground flashes. The average claim is near $1000 with the more significant damage to structures resulting in claims up to the $100 000 range. Overall, about 5 percent of all insurance claims involve lightning damage, but that percentage approaches 50 percent in Florida during the summer months. The National Fire Protection Association (NFPA), the organization that publishes the US lightning protection standard NFPA 780:2004, reports that there are about 30 000 lightning-caused house fires in the United States each year, with an annual cost of about $175 million. Interestingly, about 30 percent of all church fires are lightning-related, probably because the height of church steeples makes them preferred lightning targets.

During the summer of 1999, lightning ignited more than 2000 forest fires in Florida alone. Suppression costs for those fires totaled $160 million, and property losses amounted to almost $400 million including 126 homes, 25 businesses, and 86 vehicles. About half of the roughly 20 000 wildfires that occur annually in the western United States are caused by lightning. The US Forest Service reported that more than 15 000 square miles (about 10 million acres) burned in the lower 48 states in the fiscal year ending September 30, 2006, the largest yearly burned area since reliable record-keeping began in 1960 (for an up-to-date report, see www.ncdc.noaa.gov/oa/climate/research/monitoring.html). The firefighting cost for fiscal 2006 was a record $1.5 billion. The average annual burned area for the 10 years prior to 2006 was about 5 million acres, with a comparably smaller average firefighting cost than for fiscal 2006, although 8.6 million acres burned in 2005 with a cost of about $700 million. Major forest fires in the western United States have been more frequent and destructive during the past 20 years compared with the previous 20 years, a situation that has been attributed to the rising average temperature in the region producing a longer fire season with earlier spring snowmelts and more lightning. The summer temperature increase during this period has been between 0.5 and 1.0 °C in the western United States. Further, it has been estimated from different studies that the global surface temperature is likely to rise between 2 and 5 °C by the end of the twenty-first century (IPCC 2001), the recent and predicted annual temperature rise being about 0.25 °C per decade, with a resultant increase in convection, thunderstorms, and lightning (Williams 1992, 1994, 1999, Price 2000).

In summers with little precipitation, forests are particularly susceptible to the ignition and spread of lightning-caused fires. One of those years was 1988. In an average year, lightning ignites about 15 fires in Yellowstone Park, located primarily in Wyoming, but also extending into Idaho and Montana. Most such Yellowstone fires are small, and only about 5 percent burn more than 100 acres. In the early summer of 1988, lightning started 20 fires in the park, the first occurring on June 22 in a small group of pines. Subsequently, additional fires were ignited by both lightning and careless individuals. Almost no rain fell in Yellowstone during the summer of 1988. By mid-July, over 9500 firefighters, including four US Army battalions, were actively working to contain the various fires burning in the greater Yellowstone area. On August 20, known as "Black Saturday," a cold front passed through Yellowstone with sustained winds of 30 to 40 miles per hours and gusts up to 70 miles per hour, transforming the fires into firestorms with flames leaping as high as 200 feet. Overall, about 25 000 firefighters battled the fires which continued until the winter's first snows fell. About $150 million was eventually spent on fire suppression. Roughly 36 percent of Yellowstone Park proper, about 800 000 acres, was burned; and a total of 1.4 million acres inside and surrounding the park were consumed by fire. In the conflagration, 67 structures were destroyed with an estimated property damage of more than $3 million. Many animals were also killed, including 345 elk. Some perspective on Yellowstone fires and lightning is found in Renkin and Despain (1989, 1992), and Romme *et al.* (1995).

The National Board of Fire Underwriters reports that lightning is the primary cause of fires on farms, resulting in the loss of many millions of dollars of farm buildings and equipment annually. Lightning is also responsible for more than 80 percent of all livestock losses.

Lightning is responsible for about $2 billion annually in commercial airline operating cost and passenger delay, according to an estimate from the National Oceanic and Atmospheric Administration (NOAA). From 1988 to 1996, the US Air Force reported direct repair costs of about $1.6 billion resulting from lightning damage to military aircraft. About half of all weather-related in-flight accidents to military aircraft are thought to be caused by lightning.

The journal *Weatherwise* (Schlattler 2006), in its Weatherqueries column, answered the question, "What is the worst case of damage caused by lightning?" in part as follows.

In Brescia, Italy, on August 18, 1769, lightning struck the Church of San Nazaro, where 200,000 pounds of gun-powder, belonging to the Republic of Venice, had been stored. The resulting explosion killed 3,000 people and destroyed a sixth of the city.

On the showery afternoon of March 26, 1987, an Atlas-Centaur 67 vehicle carrying a naval communication satellite was launched from Cape Canaveral, Florida. As the spacecraft neared 12,000 feet, it triggered a lightning strike that upset the vehicle guidance system. The spacecraft went into an unplanned rotation, started to break apart, and had to be destroyed. The dollar loss of the rocket and payload amounted to $191 million.

The National Lightning Safety Institute (NLSI) notes that the most expensive U.S. civilian lightning incident was to a Denver warehouse that was struck on July 23, 1997. Losses to the building and contents totaled more than $50 million. NLSI estimates that annual lightning losses in the United States are as great as $4–5 billion.

The Atlas-Centaur 67 case, as well as other reports of lightning accidents involving airships, airplanes, and launch vehicles, is further considered in Section 9.3. Additional historical cases of lightning damage are found in Section 10.1 and Section 14.1.

A computer newsletter has estimated that lightning causes damage to about 100 000 computers per year with associated costs that might well approach or even exceed $100 million.

The various unreferenced damage estimates and costs found in this section were compiled from a variety of sources including newspaper and magazine articles the author has collected, the World Wide Web, and the National Lightning Safety Institute. Some estimates, referenced or unreferenced, may not be very accurate, but, overall, the various estimates provide a reasonable view of the cost of lightning damage.

2.2   Conductors and insulators

All materials conduct electricity to some extent. It is usual to label as "insulators" those materials that do not conduct electricity very well, such as wood and bricks, and to label as "conductors" materials that do conduct electricity well, especially metals such as copper and aluminum. A conductor of sufficient size will allow the current of a total lightning flash to flow in it without significantly increasing the temperature of the conductor. For example, a copper wire that is 1 cm in diameter (0.4 inches) passes this test for even the most severe lightning. A wire diameter that is 20 percent or so smaller is recommended by various standards (see Section 4.4). When lightning current is injected into an insulator, which attempts to resist that flow of current, considerable heat is generated in the material by the "resistance" to the current flow, as we shall discuss in the next section. Often, as in the example of a tree (see Chapter 11), that heat transforms the water or other material within the insulator to high-pressure steam causing the insulator to fracture or explode. Conductors also have electrical resistance, but their resistance is relatively small compared with that of insulators.

2.3   Lightning characteristics pertinent to damage

The damage that an object suffers in being struck by lightning depends on both the characteristics of the lightning and the properties of the object, particularly the ability or inability of the object to conduct electricity and to dissipate heat. The most interesting physical characteristics of lightning are the various properties of the time-varying current and of the radio frequency (RF) electromagnetic fields (causing the static you hear on your AM radio and the snow you see on your TV when lightning occurs nearby). Damage, such as shattered window glass, can also be caused by the acoustic shock wave from very close lightning, that is, from the high air pressure associated with very close thunder, although this is not a common occurrence. This free-air acoustic shock wave can be considerably enhanced if lightning current penetrates and explodes an insulator of significant size such as a tree (see Section 11.3).

Four properties of the lightning current waveform can be related to the most important lightning-caused damage. These four properties are identified in the hand-drawn current waveforms of Fig. 2.1 for a typical first and a typical subsequent return stroke lowering negative charge to ground (see Section 1.3). Definitions of the most common parameters of the current are illustrated in Fig. 2.2. Actual measured current waveforms are presented in Figs. 1.8, 1.9, and 2.2, and severe current waveforms specified for lightning testing are found in Fig. 9.10. Table 2.1 shows the ranges of various parameters associated with the

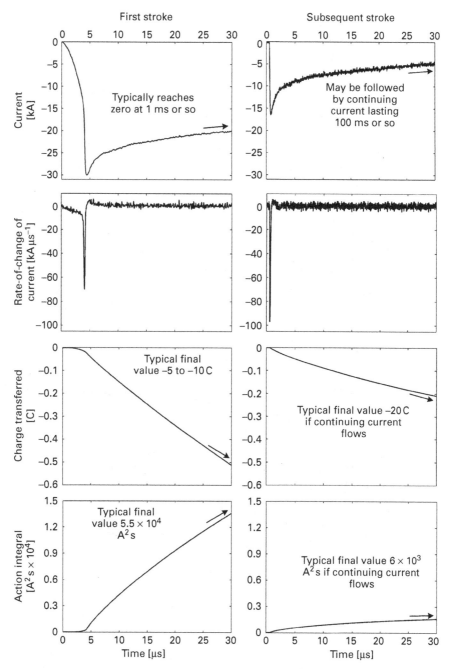

Fig. 2.1 Features of the current waveform of a lightning first and subsequent return stroke transferring negative charge to ground. These are hand-drawn curves representing typical values found in the lightning literature.

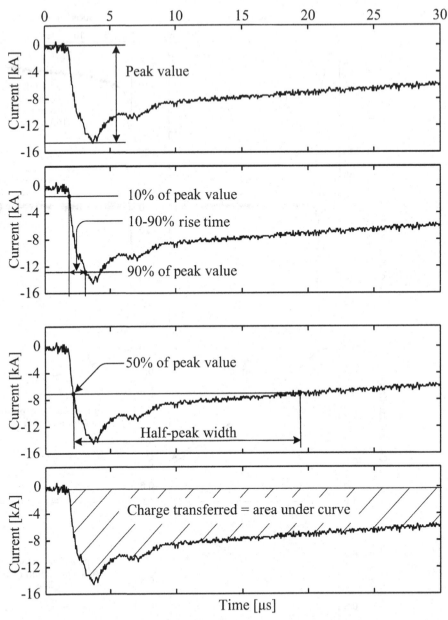

Fig. 2.2   Definitions of the most common return stroke parameters: peak value, 10–90 percent risetime, half-peak width, and charge transferred (area under current vs. time curve). The current waveform is that of a triggered lightning stroke (see Section 13.2). Triggered strokes are thought to be similar, if not identical, to subsequent strokes (strokes following the first stroke) in natural lightning.

Table 2.1 Lightning current parameters.

Number of events	Parameters	Unit	Percentage of cases exceeding tabulated value		
			95%	50%	5%
	Peak current (minimum 2 kA)				
101	Negative first strokes	kA	14	30	80
135	Negative subsequent strokes	kA	4.6	12	30
20	Positive first strokes (no positive subsequent strokes recorded)	kA	4.6	35	250
	Charge				
93	Negative first strokes	C	1.1	5.2	24
122	Negative subsequent strokes	C	0.2	1.4	11
94	Negative flashes	C	1.3	7.5	40
26	Positive flashes	C	20	80	350
	Impulse charge[a]				
90	Negative first strokes	C	1.1	4.5	20
117	Negative subsequent strokes	C	0.22	0.95	4.0
25	Positive first strokes	C	2.0	16	150
	Front duration (2 kA to peak)				
89	Negative first strokes	μs	1.8	5.5	18
118	Negative subsequent strokes	μs	0.22	1.1	4.5
19	Positive first stroke	μs	3.5	22	200
	Maximum di/dt[b]				
92	Negative first strokes	kA μs^{-1}	5.5	12	32
122	Negative subsequent strokes	kA μs^{-1}	12	40	120
21	Positive first strokes	kA μs^{-1}	0.20	2.4	32
	Stroke duration (2 kA to half-value)				
90	Negative first strokes	μs	30	75	200
115	Negative subsequent strokes	μs	6.5	32	140
16	Positive first strokes	μs	25	230	2000
	Action integral ($\int i^2 dt$)				
91	Negative first strokes	A^{2}s	6.0×10^3	5.5×10^4	5.5×10^5
88	Negative subsequent strokes	A^{2}s	5.5×10^2	6.0×10^3	5.2×10^4
26	Positive first strokes	A^{2}s	2.5×10^4	6.5×10^3	1.5×10^7
	Time interval				
133	Between negative strokes	ms	7	33	150
	Flash duration				
94	Negative (including single-stroke flashes)	ms	0.15	13	1100
39	Negative (excluding single-stroke flashes)	ms	31	180	900
24	Positives (single-stroke flashes only)	ms	14	85	500

[a] Impulse charge is the charge contained in the rapidly changing part of the return stroke waveform. It is somewhat subjective.
[b] Maximum current derivative is likely to be underestimated because measurements were made by photographing oscilloscope traces of finite width. Typical values for first and subsequent strokes to small, well-grounded objects are thought to be near 100 kA μs^{-1}.
Adapted from Berger *et al.* (1975).

Table 2.2 Severe values of lightning current parameters for (1) first strokes without continuing current, (2) subsequent strokes without continuing current, (3) continuing current, and (4) complete flashes, according to IEC 62305-1:2006. The wave shapes of the first and subsequent strokes are characterized by a front time T_1 and a time to half of peak value T_2, which is illustrated in Fig. 2.2. For an idealized wave shape, T_1 can be calculated as the 10 percent to 90 percent rise time, also illustrated in Fig. 2.2, multiplied by 1.25. Alternately, the front time T_1 is the peak current divided by the average steepness (rate of change of current during the 10 percent to 90 percent portion of the current rise to peak). The severe first stroke waveform is a 10/350 µs wave with a peak current of 200 kA; the severe subsequent stroke waveform is a 0.25/100 µs wave with a peak current of 50 kA.

	Symbol	Unit	Value
First stroke			
Peak current	I_p	kA	200
Charge	Q_s	C	100
Action integral (specific energy)	W/R	MJ Ω^{-1}	10
Front time, half-value time	T_1/T_2	µs	10/350
Subsequent stroke			
Peak current	I_p	kA	50
Average steepness	dI/dt	kA µs^{-1}	200
Front time, half-value time	T_1/T_2	µs	0.25/100
Continuing current			
Charge	Q_{cc}	C	200
Duration	T_{cc}	s	0.5
Flash			
Flash charge	Q_{flash}	C	300

lightning current, including some of those parameters illustrated in Figs. 2.1 and 2.2, from measurements made on strikes to a 55 m tower on a mountain in Switzerland (Berger *et al.* 1975; see also Fig. 1.8). Table 2.2 gives severe values for the most important lightning parameters according to the International Electrotechnical Commission, derived primarily from the same source as the data in Table 2.1, Berger *et al.* (1975). The four damage-specific properties of lightning current identified in Fig. 2.1 are: (1) the peak current; (2) the maximum rate of change with time of the current; (3) the charge transferred by the current, calculated by integrating (a calculus operation) the current over time (which equals both the area under the current vs. time curve, as shown in Fig. 2.2, and the average current multiplied by the time that the current flows); and (4) the time-integral of the current-squared, the so-called "action integral" or "specific energy" (which also equals the average current-squared multiplied by the time the current flows). We now briefly examine each of these four properties and the type of damage to which they are related.

2.3.1   Peak current

For objects or systems that appear as a "resistive" impedance to lightning current flow (such as, under most conditions, a ground rod driven into the Earth or a long power line), the voltage $V(t)$ with respect to remote ground will be proportional to the current $I(t)$. Expressed mathematically via Ohm's Law, $V(t) = R \times I(t)$, where $V(t)$ is a function of time and is measured in volts, current I is in amperes, and the effective resistance at the strike point, R, is in ohms. Hence, the peak voltage V_p is equal to R multiplied by the peak current I_p. For example, assume that a typical first stroke peak current of 30 kA (Fig. 2.1, Table 2.1) is injected into a power-line phase conductor. Typical power lines have an inherent resistive "characteristic impedance" of about 500 ohms. The effective resistance of the line at the lightning strike point is 250 ohms since 500 ohms is "seen" in each direction from the strike point (the details of this calculation are unimportant to the general argument). Without the presence of lightning arresters to limit the voltage, the lightning peak current produces a peak line voltage of 7.5×10^6 volts (250 ohms × 30 kA) with respect to the Earth, as discussed further in Section 12.2. Such a large voltage will cause an electric discharge from the conductor that is struck by lightning to adjacent phase or neutral conductors or to ground across insulating materials or through the air. The breakdown voltage for a 1 m non-uniform air gap in the laboratory (e.g., vertical rod to horizontal plane) when the rod is negative (like most lightning), and the rise time of the voltage is about a microsecond, is about 500 kV (see Section 3.3). Thus 7.5×10^6 volts should be able to produce an electrical discharge (spark, arc) through the air of roughly 15 m (7.5×10^6 volts ÷ 500 kV m^{-1}). On a typical neighborhood power line (called a distribution line), the breakdown voltage (insulation level) between the various phase and neutral conductors is 100 to 300 kV (see Section 12.2). Transmission lines have a considerably higher insulation strength because they are designed to operate at higher voltages, typically hundreds of thousands of volts. Figure 2.3 shows a rare photograph of a lightning-caused power-line flashover. The flashover may have been induced by the downward-branched first stroke (seen in the photograph) which apparently did not strike the line or (and) may have been caused by one or more subsequent strokes outside the camera view which did directly strike the line. The hand-held camera moved slightly during the time exposure, leading to the appearance of two poles, each illuminated by a different stroke in the flash. In actuality, there was only one pole. Two flashovers are also evident, by virtue of the camera motion. Note the wood splinters ejected from the pole by the flashover between the single phase wire and the neutral below as well as wood ejected further down the pole.

2.3.2   Maximum rate-of-change of current

For objects that present an "inductive" impedance, such as wires in an electronic system or down conductors in a lightning protection system, the voltage $V(t)$ across a length of wire will be proportional to the rate-of-change of the lightning current

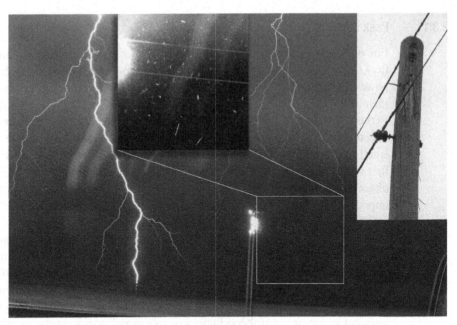

Fig. 2.3   Photograph of a lightning-caused flashover on a 12 kV single-phase power line near Shiloh, Tennessee. Photograph by Bryan S. Gross.

with respect to time, $dI(t)/dt$, in the wire; that is, $V(t) = L\, dI(t)/dt$, where L is the inductance of the length of wire. Thus, the peak inductive voltage is proportional to the maximum rate-of-change of current, which generally occurs during the current's initial rise to peak value (Fig. 2.1). For example, as illustrated in Fig. 2.4, assume that a "bonding" wire connecting two electronic systems in a structure (for example, the telephone ground and the power ground) has an inductance per unit length of 10^{-6} henries per meter ($H\,m^{-1}$) and that 1 percent of the typical maximum rate of change of the subsequent stroke current of $10^{11}\,A\,s^{-1}$ ($100\,kA\,\mu s^{-1}$) flows in the bonding wire producing $dI/dt = 10^{9}$ amperes per second in the wire. Under these circumstances a voltage of 1000 volts will be produced across each meter of the wire ($10^{-6}\,H\,m^{-1} \times 10^{9}\,A\,s^{-1}$). In the example of Fig. 2.4, the wire connecting (bonding) the telephone ground rod and the power ground rod is located outside the house, as is common, where it can carry a fraction of the lightning current by virtue of a strike to the power line, to the telephone line, to the ground near the house, or to the house. The resultant voltage difference impressed on an electronic circuit board to which both the telephone and power grounds are attached may well cause damage to the board and its components. Similarly, an individual talking on the telephone while leaning against a refrigerator whose case is bonded to the power ground may receive a severe shock, or worse. It is easy to understand how even a very small fraction of the rapidly changing lightning current circulating in various grounding and bonding wires can destroy solid-state electronic circuits, since transistors can be rendered inoperable by voltages as small as tens of volts.

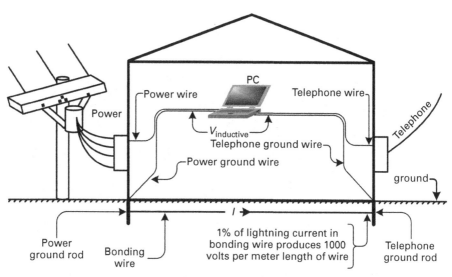

Fig. 2.4   Illustration showing how deleterious inductive voltages can be produced by a lightning strike on or near a structure with physically separated but electrically bonded grounding points for power and telephone services.

Sparks associated with inductive voltages can ignite flammable materials, an impressive example of which is given in Fig. 2.5a,b. The explosives storage bunker shown in Fig. 2.5a was protected with a standard external lightning protection system. It is thought that voltage differences between unconnected metal reinforcing rods (rebar) in the roof and sidewalls of the bunker caused sparks in the interior of the structure that ignited the stored explosives. Inductive voltage drops in the external lightning protection system apparently led to sparking between the various pieces of rebar. High interior electric fields would also have been generated. The explosion, which occurred in Austria, killed one individual. Additional consideration of inductive voltages, including a discussion of inductive "side-flashes," is found in Section 4.4.

Note that in Table 2.2 only the rate of change of current from subsequent strokes is listed because Berger *et al.* (1975) found the value of that parameter for first strokes to be considerably lower than for subsequent strokes, as indicated in Table 2.1. Recent measurements show that the latter portion of the first stroke current rise to peak is similar to the subsequent stroke overall current rise to peak, and hence that the peak value of the first and subsequent stroke current rate of change is similar, as illustrated in Fig. 2.1.

2.3.3   Charge transfer

The total charge transfer is the integral of the current over the time that current flows (the area under the current vs. time curve) or, equivalently, the average

Fig. 2.5 A concrete explosives storage facility that was struck by lightning, (a) before and (b) after. Courtesy of Marvin Morris.

current multiplied by the duration of current flow. The charge transfer that occurs by a given time is the integral of the current to that time. The severity of heating or burn-through of metal sheets such as airplane wing surfaces (see Figs. 9.7–9.9) and metal roofs is, to a first approximation, proportional to the lightning charge transferred; which is in turn proportional to the energy delivered to the surface. This is the case because the input power $P(t)$ to the conductor surface is the product of the current $I(t)$ and the relatively constant voltage difference V between the lightning discharge (arc) and the metal at the arc–metal interface, this voltage difference generally being of the order of 10 volts. The energy $E(t)$ delivered to the surface in time t is the time-integral of the power $P(t)$, that is,

$$E(t) = \int_0^t VI(t')dt' \approx VQ(t) \tag{2.1}$$

the roughly constant voltage drop V of the order of 10 volts multiplied by Q, the charge transferred up to time t. Generally, large charge transfers are due to long-duration (tens to hundreds of milliseconds) lightning currents, such as continuing currents between strokes, whose magnitude is in the 100 to 1000 ampere range, rather than to return strokes which have larger amplitude but shorter duration currents and hence produce relatively small charge transfers. Typical total charge transfers for lightning discharges are 30 coulombs or so. The heating or burn-through of metal sheets depends not only on the energy delivered to a point on the metal by the lightning current but also on the metal's thermal conductivity and its thickness, that is, its ability to carry heat away from the strike point while the lightning is occurring. For example, about 30 coulombs of charge delivered to the surface of a 3 millimeter thick (about 1/8 inch) plate of aluminum in 0.1 second will just burn through the plate, whereas it will require about 100 coulombs to burn through the plate if the charge transfer takes place in 0.5 seconds since there is sufficient time for heat to flow away from the strike point. Thinner plates require less charge for burn-through. Half the aluminum thickness requires about half the charge to burn through for the same time-duration of charge input. The damage done to a stationary metal surface by a typical lightning charge transfer is similar to that done by an arc welder delivering a few hundred amperes of current to the metal surface for a fraction of a second. It is interesting to note that lightning strikes to residential and commercial structures sometimes result in small holes being burned in copper water-piping, producing leaks. More details on the heating and burn-through of metal sheets of various materials and thicknesses are found in Bellaschi (1941), McEachron and Hagenguth (1942), Hagenguth (1949), Brick (1968), Kofoid (1970), and Testé *et al.* (2000).

2.3.4 Action integral (specific energy)

The heating and melting of resistive materials in which lightning current flows and the explosion of poorly conducting materials (insulators) are, to a first approximation, related to the value of the action integral (also known as the specific energy); that is, the time integral of the Joule power dissipated, $V(t) \times I(t) = I(t)^2 \times R$, since $V(t) = R \times I(t)$, for the special case that $R = 1$ ohm. Thus the action integral can be written as

$$\int_0^t I^2(t')dt' \tag{2.2}$$

The action integral is a measure of the heat generated by lightning current in a strike object characterized by a resistance of $R = 1$ ohm. The heat generated in any particular struck object is found by multiplying the action integral by the value of the resistance of that object. About 5 percent of negative first strokes in ground flashes have action integrals exceeding 5.5×10^5 amperes-squared times seconds (A^2s); about 5 percent of positive strokes have action integrals exceeding $1.5 \times 10^7 A^2$s (see Table 2.1, Table 2.2, Section 4.4). In the case of most poorly conducting materials, the heat represented by the action integral vaporizes the internal material and the resultant gas pressure causes an explosive fracture. Examples are shown in Fig. 9.6, a photograph of lightning damage to the radome of an aircraft, and in the photographs of lightning damage to trees of Figs. 11.4–11.8. An additional example, that of a house burned to the ground by lightning, before and after, is shown in Fig. 2.6a and b. Houses struck by lightning sometimes exhibit split structural wood without burning, as do trees, and nails that are pulled (or pushed) from the wood, apparently by the generation of steam at the tip of the imbedded nail. Another common type of damage to residential structures related to the action integral is the bursting of light bulbs, often accompanied by damaged and sooty electrical outlets. In addition to heating effects, the action integral is also a measure of some mechanical effects, such as the ability of the lightning current to crush hollow metal tubes (as found, for example, as structural components in some aircraft) through which the current flows. The crushing effect (which arises from the magnetic field created by the flowing current) is both a function of the instantaneous force, which is proportional to the square of the current, and the time that the force is applied. To crush a hollow metal tube, the applied force must also exceed some threshold value.

Lightning electromagnetic (electric and magnetic) fields that impinge on any conducting object induce currents and voltage in that object. Two properties of the electromagnetic fields are sufficient to describe most of the important damaging effects, commonly the destruction of electronic components on electronic circuit boards: (1) the peak values of the electric and magnetic fields and (2) the maximum rate of change of those fields with respect to time. For certain types of unintended antennas such as elevated conductors that are "capacitively" coupled to ground, the peak voltage induced on the conductors (with respect to ground) is proportional to the peak electric field. For other unintended antennas such as loops of wire in electronic circuits, some underground communication cables, and elevated conductors that are resistively coupled to ground, the peak induced voltage is proportional to the maximum rate of change of the electric or the magnetic field. The degree of coupling of fields through apertures (non-metallic openings, like windows) in the metal skins of aircraft and spacecraft is generally proportional to the rate of change of the electric and magnetic fields.

Perhaps the most common deleterious electromagnetic coupling is that of the lightning magnetic field to loops of wire, often in electronic circuits. By Faraday's Law, the magnitude of the voltage induced in a loop of wire, which can then appear, for example, across a transistor attached to that loop, is proportional to the component of the rate of change of the magnetic flux density $\mathbf{B}(t)$ that is perpendicular (normal) to the plane of the loop multiplied by the area A of the loop (e.g., Sadiku 2007):

Fig. 2.6   (a) A house approaching the final stages of construction in Albuquerque, New Mexico. (b) The remains of the house after a lightning-ignited fire. The scaffolding shown in (a) was located inside the house when the lightning struck. Photographs by Tom Bretz.

$$|V_{\text{loop}}| = A \cdot \mathrm{d}B_{\text{n}}(t)/\mathrm{d}t, \tag{2.3}$$

where B_{n} is the normal component, so long as the magnetic flux density is roughly constant across the loop. Faraday's Law without (and with) that assumption is given and is illustrated in Fig. 2.7. For example, 10 m from a down conductor carrying a lightning current $I(t)$ along the side of a lightning-protected structure, as

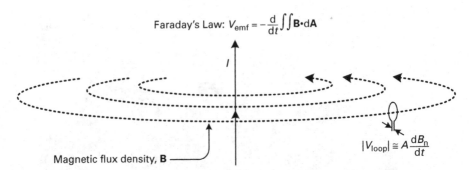

Fig. 2.7   Illustration of the magnetic flux density surrounding a long, current-carrying conductor, Faraday's Law, and the magnitude of voltage induced in a loop of area A located in the magnetic field.

illustrated in Fig. 2.8, the magnetic flux density (assumed to be from a very long current-carrying wire) is well approximated by

$$B(t) = \frac{\mu_0 I(t)}{2\pi r} = \frac{4\pi \times 10^{-7} I(t)}{2\pi \times 10} = 2 \times 10^{-8} I(t) \ \text{webers m}^{-2} \tag{2.4}$$

where r is the distance from the lightning current to the loop and μ_0 is the permeability of free space. Hence for a peak current rate of change $dI(t)/dt$ of $10^{11} \ \text{A s}^{-1}$, the rate-of-change of $B(t)$ is

$$\frac{dB(t)}{dt} = 2 \times 10^{-8} \frac{dI(t)}{dt} = 2 \times 10^3 \ \text{webers m}^{-2} \text{s}^{-1} \tag{2.5}$$

Thus, for a loop with a small break, an area of $10^{-2} \, \text{m}^2$ (a square loop 10 cm on a side as might be found on a printed circuit board), and the magnetic flux density normal to the loop from lightning current 10 m away, there will be 20 V across the ends of the loop. For the same lightning current, either a closer or a larger loop will have a larger induced voltage. All power and communication wiring in a structure will present loops whose areas are likely to be measured in square meters. A reduction in the magnetic flux density inside a structure due to the lightning current in a single down conductor on the outside of the structure can be achieved if many separated down conductors are allowed to share the lightning current. A similar effect is achieved if the current is distributed over the structure's side walls, as can be the case for a structure with walls of concrete containing connected (bonded) metal reinforcing rods (rebar). As a simple example of this effect, the total magnetic field midway between two down conductors carrying equal current will be zero since the magnetic fields produced by the two individual currents are exactly equal and in opposite directions at that point. Structures with multiple down conductors or with somewhat conducting walls act as imperfect Faraday cages (see Section 3.1).

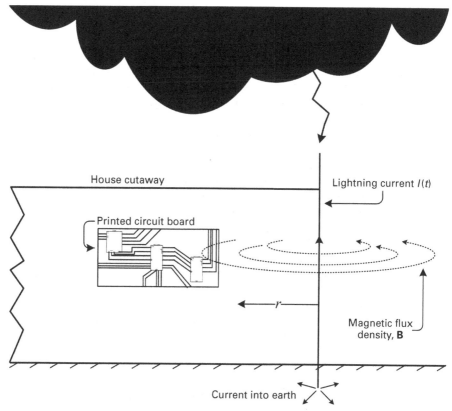

Fig. 2.8 Illustration of magnetic coupling to a circuit board from lightning current in the down conductor of a lightning protection system.

A discussion of the shielding effects in such structures including theory, model measurements, and reference to international standards, is found in Metwally *et al.* (2004, 2006) and in Zischank *et al.* (2004).

References

Bellaschi, P. L. 1941. Lightning strokes in field and laboratory III. *AIEE Trans.* **60**: 1248–1256.

Berger, K., Anderson, R. B., and Kroninger, H. 1975. Parameters of lightning flashes. *Electra* **80**: 23–37.

Brick, R. O. 1968. A method for establishing lightning-resistant skin thickness requirements for aircraft. *Lightning and Static Electricity Conference*, AFAL-TR-**68**-290, Part II, 295–317.

Curran, E. B., Holle, R. L. and Lopez, R. E. 2000. Lightning casualties and damages in the United States from 1959 to 1994. *J. Climate* **13**: 3448–3464.

Hagenguth, J. H. 1949. Lightning stroke damage to aircraft. *AIEE Trans.* **68**(II): 1036–1046.

Holle, R. L., López, R. E., Arnold, L. J. and Endres, J. 1996. Insured lightning-caused property damage in three western states. *J. Appl. Meteorol.* **35**: 1344–1351.

IPCC 2001. *Climate Change 2001: The Scientific Basis. Contribution of Working Group I to the Third Assessment Report of the Intergovernmental Panel on Climate Change*, ed. J. T. Houghton, Y. Ding, D. J. Griggs *et al.* Cambridge: Cambridge University Press.

Kofoid, M. J. 1970. Lightning discharge heating of aircraft skins. *J. Aircraft* **1**: 21–26.

McEachron, K. B. and Hagenguth, J. H. 1942. Effects of lightning on thin metal skins. *AIEE Trans.* **61**: 559–64.

Metwally, I. A., Zischank, W. J. and Heidler, F. H. 2004. Measurement of magnetic fields inside single-and-double-layer reinforced concrete buildings during simulated lightning currents. *IEEE Trans. Electromagn. Compatibility* **46**: 208–221.

Metwally, I. A., Heidler, F. H. and Zischank, W. J. 2006. Magnetic fields and loop voltages inside reduced and full-scale structures produced by direct lightning strikes. *IEEE Trans. Electromagn. Compatibility* **48**: 414–426.

NFPA 780: 2004. *Standard for the Installation of Lightning Protection Systems*. Quincy, MA: National Five Protection Association.

Price, C. 2000. Evidence for a link between global lightning activity and upper tropospheric water vapour. *Nature* **406**: 290–293.

Renkin, R. A. and Despain, D. G. 1989. Historical perspective on the Yellowstone Fires of 1988. *Bioscience* **39**: 696–699.

Renkin, R. A. and Despain, D. G. 1992. Fuel moisture, forest type, and lightning-caused fire in Yellowstone National Park. *Can. J. Forest Res.* **22**: 37–45.

Romme, W. H., Turner, M. G., Wallace, L. L. and Walker, J. S. 1995. Aspen elk and fire in northern Yellowstone National Park. *Ecology* **76**: 2097–2106.

Sadiku, M. 2007. *Element of Electromagnetics*, 4th edition. Oxford: Oxford University Press.

Schlattler, T. 2006. Weather queries. *Weatherwise* **59** (2).

Sugarman, R. 1978. New York City's blackout: a $350 million drain. *IEEE Spectrum* **15**: 44–46.

Testé, Ph., Leblanc, T., Uhlig, F. and Chabrerie, J.-P. 2000. 3D modeling of the heating of a metal sheet by a moving arc: application to aircraft lightning protection. *Eur. Phys. J.* **AP11**: 197–204.

Williams, E. R. 1992. The Schumann resonance: a global tropical thermometer. *Science* **256**: 1184–1187.

Williams, E. R. 1994. Global circuit response to seasonal variations in global surface air temperature. *Mon. Weather Rev.* **122**: 1917–1919.

Williams, E. R. 1999. Global circuit response to temperature on distinct time scales: a status report. In *Atmospheric and Ionospheric Electromagnetic Phenomena Associated with Earthquakes*, ed. M. Hayakawa. Tokyo: Terra Scientific Publishing Company (TERRAPUB), pp. 939–949.

Wilson, G. L., and Zarakas, P. 1978. Anatomy of a blackout: how's and why's of the series of events that led to the shutdown of New York's power in July 1977. *IEEE Spectrum* **15**: 39–46.

Zischank, W., Heidler, F., Wiesinger, J. *et al.* 2004. Laboratory simulation of direct lightning strokes to a modeled building: measurement of magnetic fields and induced voltages. *J. Electrostat.* **60**: 223–232.

3   General methods for lightning protection: Faraday cages, topological shields; and more practical approaches: cone of protection and rolling sphere methods

3.1   Is there perfect protection?

Essentially perfect protection from both the direct currents and the electric and magnetic fields of lightning can be found inside a closed metal structure with an appropriate wall thickness and with no holes or openings (apertures) in the walls, including openings associated with wall penetrations by metallic conductors such as those carrying power and communication signals. Such a closed conducting structure is called an electrodynamic Faraday cage, or simply a Faraday cage or Faraday shield. But what is an appropriate wall thickness? According to electromagnetic field theory (as described by the four famous Maxwell's Equations), electrical currents and electromagnetic fields are maximum on the outside surface of closed conductors and, for a signal of a given frequency, decrease inward to about one-third (actually $1/e$ where $e = 2.718$) of their values at the outer surface for each "skin depth." That is, at two skin depths, a single-frequency signal is decreased to about one-tenth; at four skin depths, about one-hundredth. Thus, for a shield to be optimum, the wall thickness should be many skin depths for all significant frequencies composing the lightning signal so that no appreciable signal reaches the interior of the shield. Skin depth δ is a function of the electrical resistivity ρ (or its reciprocal, the conductivity σ) of the shield material and the frequency f of the undesired signal as follows:

$$\delta = \sqrt{\rho / \pi f \mu} \qquad (3.1)$$

where μ is the magnetic permeability of the material (e.g., Sadiku 2007). Equation 3.1 is plotted in Fig. 3.1 for copper for applied currents or fields of different frequencies between power frequency (60 Hz) and 100 MHz, a few kilohertz to a few tens of megahertz being roughly the range of frequencies that are significant in direct lightning currents and in the radiated fields. Thus a copper shield several millimeters thick would usually be sufficient to protect from lightning effects, except for possible damage to the shield at the lightning–copper interface, as discussed in Section 2.3, if the lightning current strikes the shield directly. Faraday shields are

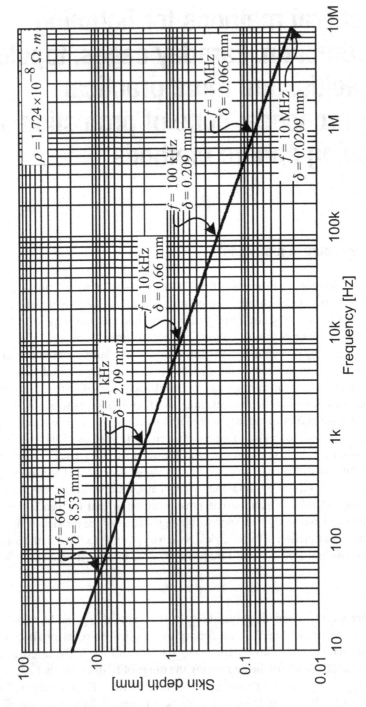

Fig. 3.1   Skin depth vs. frequency for copper, plotted on a log–log scale.

sometimes constructed from two layers of copper screen on a wood frame, forming a so-called "screen room" in which battery-operated scientific experiments can be performed without interference from externally generated electromagnetic noise such as from lightning, radio or TV stations, or light switch operation.

The concept of perfect lightning protection afforded by a Faraday cage is, unfortunately, of limited practical value. Most structures in the real world have metallic penetrations for plumbing, electricity, and communications, not to mention windows. However, a systematic approach to lightning protection has been developed that combines the general principle that time-varying currents and electromagnetic fields are seriously attenuated when propagating into the interior of a closed, conducting cage with the use of voltage-limiting surge protective devices (SPDs) to control the magnitude of signals entering the cage via external conductors. This approach is known as "topological shielding with transient protection" (e.g., Tesche 1978, Vance 1980), and it allows an optimal lightning protection system to be designed for most structures and their contents. The technique consists of nesting solid metal or metal-screen cages (shields) within each other, as illustrated in Fig. 3.2. The outside of each shield is electrically connected ("grounded") to the inside of the shield enclosing it, with the outermost shield being grounded to the Earth, as also shown in Fig. 3.2. Achieving a low value of Earth ground is not particularly important to the effectiveness of the overall system. No separate ground wires are allowed to penetrate through the individual shields, a critical requirement in the topological shielding scheme. Conducting wires entering a given shield must pass through a surge protective device whose ground lead is connected to the outside of that shield. Therefore, at each successively inner shield, the surge protection device reduces deleterious electrical power and electromagnetic field levels both on incoming wires and in space. The proper coordination of the characteristics of multiple surge protection devices in series is discussed in Section 6.6. In the real world, the outermost shield could be, for example, an all-aluminum building or the steel structure of a battleship. Nested inside that outer shield could be a large screen room, and inside that, metal cabinets containing computers or other sensitive electronics. If properly configured, such a system should be able to withstand the current of a direct lightning strike, illumination by a high power radar, or any other potentially severe currents or electromagnetic fields.

The effectiveness of any lightning protection system can be assessed by determining how closely it approximates a topologically shielded and surge protected system.

3.2   Practical protection: types of protection

Two factors must be considered in ordinary lightning protection: (1) the diversion of the lightning current away from the protected structure, which is primarily to protect the structure but also serves (although not generally recognized as such) to reduce the lightning electric and magnetic fields within the structure, to the extent that the diversion wires effectively form a reasonable approximation to a Faraday

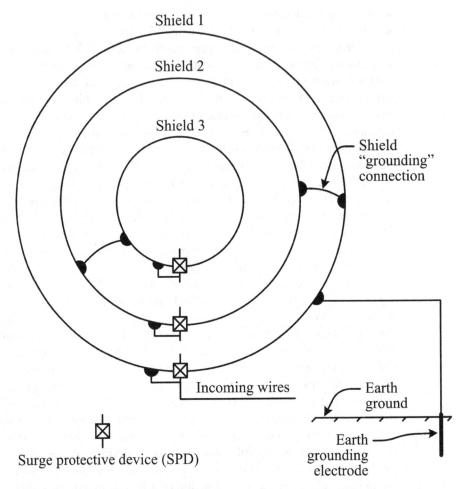

Fig. 3.2   Schematic drawing of a three-shield system of topological shielding with transient protection to provide optimum protection from the deleterious effect of lightning currents and fields. The highest degree of protection is found inside shield 3. Even higher degrees of protection can be achieved by nesting additional shields within shield 3.

shield; and (2) the limiting of currents and voltages on power and communication systems via surge protective devices. As we have noted in Section 3.1, these two aspects of protection are related; however, we present them separately here, the way they are generally treated in codes and standards.

We look first at current diversion. Lightning current is diverted away from a protected structure and ultimately into the Earth by the three electrically connected components of the protection system: (1) air terminals, which may be vertical lightning rods (also referred to as Franklin rods) connected together on the roof of the structure, or a mesh of horizontal wires on the roof, or overhead catenary wires above the roof, or a metal roof, all intended for the same purpose, to intercept the descending lightning stepped leader (see Section 1.3); (2) down conductors to

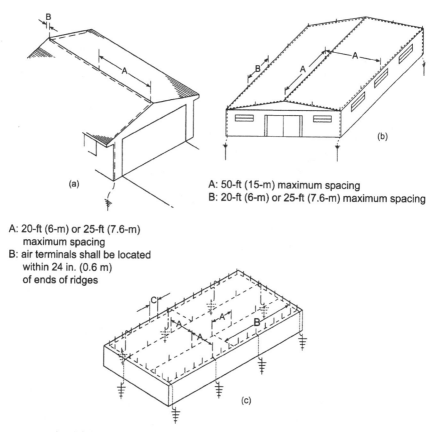

A: 20-ft (6-m) or 25-ft (7.6-m)
   maximum spacing
B: air terminals shall be located
   within 24 in. (0.6 m)
   of ends of ridges

A: 50-ft (15-m) maximum spacing
B: 20-ft (6-m) or 25-ft (7.6-m) maximum spacing

A: 50-ft (15-m) maximum spacing between air terminals
B: 150-ft (45-m) maximum length of cross run conductor permitted
without a connection from the cross run conductor to the main
perimeter or down conductor
C: 20-ft (6-m) maximum spacings between air terminals along edge

Fig. 3.3 Lightning protection of three structures with different roof types, adapted from NFPA
780:2004. (a) Pitched roof; (b) gently sloping roof; (c) flat roof. Reprinted with permission
from NFPA 780, *Installation of Lightning Protection Systems*, Copyright ©2004, National
Fire Protection Association, Quincy, MA 02169. This reprinted material is not the complete
and official position of the National Fire Protection Association on the referenced subject
which is represented only by the standard in its entirety.

carry the lightning current to the grounding electrodes; and (3) grounding electrodes to convey the current into the Earth. This general protection scheme for
structures was originally proposed by Benjamin Franklin (see Section 4.1). The
details of such lightning protection are specified, for example, in the international
lightning protection standard IEC 62305-1, 2, 3, 4, 5:2006, in US lightning protection standard NFPA 780:2004, and in many other national standards. Examples of
lightning protection systems for three structures with different types of roofs,
adopted from NFPA 780:2004 in which the roof types are defined, are shown in
Fig. 3.3. Note that the lightning rods are electrically connected by bonding conductors, shown as dashed lines, which should be considered as part of the air

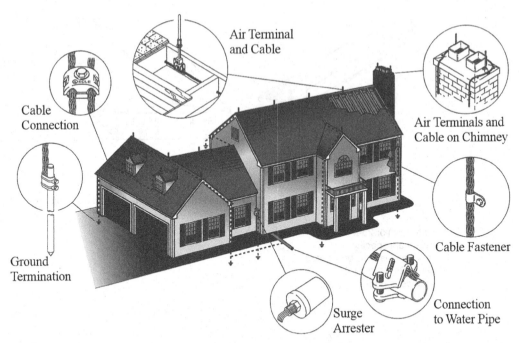

Air Terminal
and Cable

Cable
Connection

Air Terminals and
Cable on Chimney

Ground
Termination

Cable Fastener

Connection
to Water Pipe

Surge
Arrester

Fig. 3.4   Lightning protection of a residential structure. Courtesy of East Coast Lightning
Equipment, Inc.

terminal system (they may be struck although this is not the design intention), in
addition to the bonding conductors transmitting the lightning current to the down
conductors. Figure 3.4 illustrates a residential lightning protection system designed
to the specifications of NFPA 780:2004. Figure 3.5 shows similarly designed lightning protection for a multi-storied commercial building. The term Franklin rod is
sometimes used to describe only a sharp-pointed rod, consistent with Franklin's
view of the proper rod geometry (see Section 4.1; and we will discuss in Section 4.2
whether sharp rods or blunt ones are better, both being allowed by NFPA
780:2004), but is more often used to refer to any vertical lightning rod, in honor
of the inventor. If the lightning current is brought to ground by multiple down
conductors symmetrically placed around the structure (some approximation to an
all-metal building), as opposed to a single down conductor or an unsymmetrical
arrangement of down conductors, such a diversion system will also decrease the
potentially harmful effects to electronic equipment inside the structure from
induced voltages that are primarily produced by the time-varying magnetic field
whose source is the time-varying current in the down conductors (see Section 2.3).
Basically, the total magnetic field inside a symmetrical placement of multiple down
conductors is reduced by cancellation of oppositely directed magnetic fields from
the currents of individual wires. All elements of the structural lightning protection
system must be well bonded (connected together) electrically and all significant
nearby conductors, including the ground wires on incoming utilities, must be
bonded to the overall protection system to avoid, as much as possible, voltage

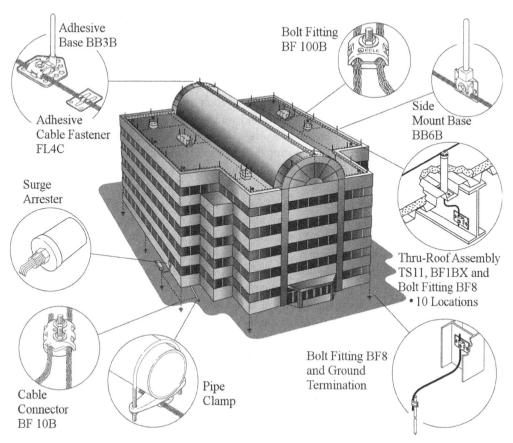

Adhesive Base BB3B

Adhesive Cable Fastener FL4C

Surge Arrester

Cable Connector BF 10B

Bolt Fitting BF 100B

Side Mount Base BB6B

Thru-Roof Assembly TS11, BF1BX and Bolt Fitting BF8 • 10 Locations

Bolt Fitting BF8 and Ground Termination

Pipe Clamp

Fig. 3.5   Lightning protection of a commercial structure. Courtesy of East Coast Lightning Equipment, Inc.

differences between the conductors that may lead to electrical breakdown between them. As noted in Section 2.3 (see Fig. 2.4), resistive voltage differences can be avoided by bonding, but inductive voltages will still be present if lightning current flows in the bonding wires. The air terminals and the down conductors of the diversion system are discussed further in Chapter 4. The grounding electrodes of the diversion system, a critical part of the structural protection system, and aspects of grounding in general, are considered in Chapter 5.

Now we survey surge protection. A more complete exposition is found in Chapter 6. The protection of electronic, power, or communication equipment within a structure should include the control of currents and voltages resulting both from direct strikes to the structure containing the equipment, and from lightning-induced current and voltage surges propagating into the structure on electric power, communication, or other metal wires and metal pipes entering the structure from outside. Signals sent into the structure on insulating (non-conducting) fiber optic links do not generally require lightning protection. Four types of current- and voltage-limiting techniques are commonly used. (1) Voltage crowbar devices limit the deleterious voltages on the protected wires to small values compared with the operating voltage and attempt to short-circuit the associated current to ground. The older carbon block arresters and the modern gas-tube arresters used by telephone companies are good examples of crowbar devices. When the voltage across such a crowbar device reaches a value of many hundreds of volts, the arrester suffers an electrical breakdown in its gas component, reducing the voltage across the arrester terminals to near zero. Silicon controlled rectifiers (SCRs) and triacs are other examples of crowbar devices. Crowbar devices do not operate instantaneously and, in general, do not operate as rapidly as the voltage clamps discussed next. (2) Voltage clamps, solid-state devices such as metal oxide varistors (MOVs), Zener and avalanche diodes, and p–n junction diodes both reflect and absorb energy while clamping the applied voltage across their terminals to a more-or-less safe value, ideally 30 to 50 percent above the system operating voltage, rather than the very small voltages allowed by crowbar devices. Voltage clamps generally can handle less energy than crowbar devices before failing. Both voltage clamps and voltage crowbar devices are referred to as surge protective devices (SPDs). (3) Circuit filters, linear electrical circuits, both reflect and absorb the frequencies that form the damaging lightning transient pulses while passing the operating waveforms, which might be communication signals of a volt or less, or 60 Hz power at 110 or 220 volts. The simplest circuit filter is a series inductor whose impedance is much higher to the frequencies comprising the unwanted transient than to the operating frequency of the electronics being protected. Special material called mu-metal may be wrapped around signal wires carrying relatively low-frequency signals. The mu-metal absorbs the higher-frequency transients but does not affect the lower frequencies. Frequently, crowbar devices, clamps, and filters are used together in a coordinated way. (4) Isolating devices such as optical isolators and isolation transformers can suppress relatively large transients. Isolators are connected in series with the equipment to be protected and represent a large series impedance to the unwanted transient signals.

The SPDs discussed above are generally connected at the input terminals of electrical devices like motors and transformers, on power plugs and signal wires of electronic components like TVs, and directly on signal and power conductors on circuit boards. The SPDs are placed between the two signal leads (if there are two) and/or between each signal lead and ground lead. As indicated in Section 3.1, in the topological shielding scheme, SPDs are also required on the outside of each level of shielding where wire penetrations exist.

3.3   The striking distance

The descending stepped leader, when it is tens to hundreds of meters above ground, will generate a relatively large electric field whose source is both the charge residing on the leader and the opposite polarity charge induced on the Earth and particularly on objects connected to the Earth but projecting above it. Such large electric fields will cause an electrical discharge (breakdown) between the tip of the stepped leader and the Earth (or an object on the Earth). For a protected structure, the object on the Earth should be lightning rods on the roof, a horizontal wire mesh laid on the roof, horizontal wires suspended above the roof by separate vertical conducting masts, or a metal roof. The physics of the "attachment process" involves the launching of an upward-connecting leader or leaders from the ground or from grounded objects such as the air terminals of a lightning protection system to connect with the downward-moving stepped leader, as discussed in Sections 1.3–1.5 (Figs. 1.6, 1.7, 1.8, and 1.12). A simple and straightforward approach to describing the situation involves calculating the electric field produced by the stepped leader charge and the induced charge on the protection system and the ground. To do so, it is first necessary to specify the charge distribution along the leader channel, generally assumed, for simplicity, to be a single vertical channel, although there may well be numerous stepped leader branches approaching ground simultaneously. Some investigators have assumed a uniform charge distribution (the same charge per meter length at any height on the idealized stepped leader channel), while others have argued that the charge per meter is largest at the bottom and decreases from the leader tip upward towards the cloud charge source, a more likely scenario from a physical point of view. Whatever the exact model assumed, the magnitude and the distribution of the leader charge with height can be estimated, partly from ground-based measurements of the leader electric field made some kilometers away, and then, given the magnitude and distribution of the leader charge, the electric field under the leader can be calculated as a function of the height of the leader tip above the ground. The "striking distance," an expression first used by Benjamin Franklin (1767), can then be defined as the distance from the leader tip to the object to be struck for which the "critical electric field" necessary to cause breakdown is reached. The critical electrical field is in turn defined as the average breakdown electric field in the gap (the gap breakdown voltage divided by the striking distance). At the specific location or locations from which upward leaders are initiated (e.g., the top of a lightning rod), the electric field must equal or exceed $3 \times 10^6 \, V \, m^{-1}$ if the location is near sea level, as this is the value of the uniform electric field that will cause electrical breakdown at standard temperature and pressure. The breakdown field is lower at higher altitudes. The final gap (striking distance) is bridged both by an upward-connecting leader from an object connected to the ground or from the ground and a downward continuation of the stepped leader. The electric field at all locations in the gap must be high enough to allow both of these discharges to propagate, an important requirement on the spatial structure of the critical electric field, in addition to being strong enough near the eventual strike point to facilitate initiation of an upward leader. It is not known what fraction of the striking distance is traversed by an upward leader (it could be most of the gap as might be inferred from Fig. 1.8) and what fraction is traversed by a downward continuation of the stepped leader. This depends on the relative average speeds of the two leaders, which may be in part determined by the character of the stepping of each. It is also not known whether the downward moving leader continues on its previous trajectory (and if so, for what distance) when the upward leader is initiated or whether it immediately deflects in the direction of the upward leader. The various representative views of the proper attachment process model according to the electric power research community are expressed in the papers of Rizk (1990, 1994a,b) which include the discussions of other researchers who question Rizk's model. The potential complexity of the situation is illustrated by Fig. 1.8 where the final gap is not traversed by a straight discharge and no stepping is evident from either the downward or upward leader after the downward leader's stepping apparently terminates at point A. Of course, the fact that no stepping is recorded on the film may just be because the step luminosity was too weak to be recorded. An equivalent definition of the striking distance to that discussed above involving the critical electric field is simply that it is the distance to the eventual strike-point from the downward-moving leader tip at the time when a successful upward-connecting leader is initiated from the eventual strike-point.

Values of the critical electric field of $500 \, kV \, m^{-1}$ for negative stepped leaders and $300 \, kV \, m^{-1}$ for positive leaders have often been assumed, based on various laboratory experiments with long sparks in non-uniform gaps (e.g., Golde 1973, 1977). These laboratory sparks typically have been generated between an elevated vertical rod and a horizontal plane below, the rod simulating the stepped leader and the plane the Earth, but there are also rod–rod data and data for more complex geometries. The criterion for breakdown in long non-uniform air gaps, including lightning, is considerably more complex than can be described by a single parameter such as the critical electric field, but such an approach has been generally used to compensate for our deficiency in understanding the physics of the attachment process and its variability. Two examples of this complexity are (1) the lack of knowledge of the detailed geometry of the region between the charged, multiply branched (umbrella of) leader steps and the variously shaped conducting structures on the ground, which determine the electric field as a function of position before the breakdown begins, and (2) the fact that the critical breakdown electric field is not a constant for a given gap geometry but depends on the rise time and the fall time of the voltage waveform applied across the gap. Voltage rise times in the $100 \, \mu s$ range are associated with waveforms called "switching surge impulses" while rise times in the microsecond range are called "steep-fronted impulses" or "lightning impulses." The exact form of the time-varying voltage applied across the final gap by the descending stepped leader is poorly known. Recently measured electric fields at ground beneath the very close step process exhibit both a microsecond-scale increase in field magnitude, presumably at the time the optical step occurs, and a subsequent steady increase in electric field during the $10 \, \mu s$ or so between steps (e.g., Dwyer *et al.* 2005). The critical electric field values given above are not inconsistent with the extrapolation of either laboratory switching impulse or steep-fronted impulse experiments to longer gaps, although the switching surge values are generally lower and sometimes are taken to be considerably lower. Reviews of the literature on long-gap electrical breakdown in the laboratory, from which extrapolation to the lightning situation has been made, are found, for example, in the book by Chowdhuri (1996), the books by Bazelyan and Raizer (1997, 2000), in the two summarizing publications of the long-spark experiments of the Les Renardieres Group (1997, 1981), and in Anderson and Tangen (1968). Note that most of the laboratory experiments discussed in these references have involved spark gaps that are smaller than a single step of the stepped leader (50 m or so above ground and about 10 m very close to ground), so one might question the validity of any extrapolation of the laboratory data to natural lightning.

The charge on the stepped leader, in addition to determining the height at which the leader produces the critical electric field, a concept which we cautiously embrace despite its inadequacies, plays an important role in determining the peak current in the return stroke. This has been inferred from measurements of the proportionality between peak current and charge transfer during strikes to a 55-m-tall tower on Mount San Salvatore in Switzerland (Berger 1972, Cooray et al. 2007), also the source of Fig. 1.8. In general, the greater the stepped leader charge, the larger the peak current of the first return stroke, a not unreasonable experimental result since it is the charge deposited on the leader channel that is released to form the return stroke current. The striking distance d is often approximated by a relatively simple formula of the form

$$d = AI_p^b \qquad\qquad (3.2)$$

where d is in meters, I_p is the return stroke peak current in kiloamperes, and A and b are constants. In Eq. (3.2), the striking distance is expressed in terms of the first stroke peak current, whose range of values is moderately well known from direct measurement (see Table 2.1), rather than in terms of the less-well-known charge on the stepped leader, and justified via the experimental link between leader charge and return stroke peak current noted above. Rizk (1990) and Cooray et al. (2007) review the literature concerned with this relationship and Eq. (3.2). Armstrong and Whitehead (1968) have suggested using $A = 6$, $b = 0.8$ in Eq. (3.2) for the case of negative lightning striking the flat Earth. This curve of peak current vs. striking distance, labeled d_1, is plotted in Fig. 3.6. If I_p is 30 kA, then from the Armstrong and Whitehead (1968) curve, the striking distance d_1 is about 90 m. IEEE Standard 998:1996 adopts $A = 8.0$ and $b = 0.65$, which for $I_p = 30$ kA yields a striking distance d_2, plotted in Fig. 3.6 as a function of peak current, of about 75 m. IEEE Standard 998:1996 notes that for five published values of A and b, in the five different studies they reference, the striking distance varies by a factor of 2 for a given current. The International Electrotechnical Commission's international lightning protection standard IEC 62305-1:2006 adopts $A = 10$ and $b = 0.65$ in calculating the relation between striking distance and peak current, a 25 percent increase in striking distance for a given peak current over the value used by IEEE Standard 998:1996. The IEC curve is plotted as d_3 in Fig. 3.6. Cooray et al. (2007) derive $A = 1.9$ and $b = 0.90$, yielding a striking distance of 41 m for $I_p = 30$ kA. The shaded area of Fig. 3.6 shows the approximate range of the majority of the calculations of striking distance to flat Earth as a function of peak current (using

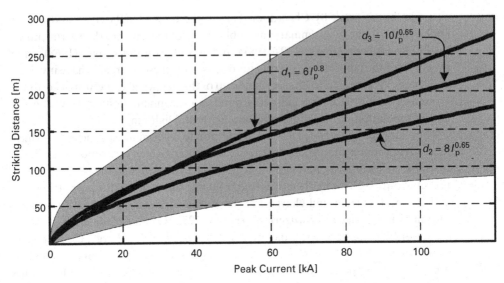

Fig. 3.6   Striking distance vs. first return stroke peak current for lightning to flat ground. Three common curves representing Eq. (3.2) with different values of the parameters A and b are plotted. The shaded region illustrates the approximate range of published calculations relating striking distance to peak current.

different input parameters and assumptions, for both positive and negative lightning) found in the lightning literature (see, for example, Golde 1973, 1977, IEEE Standard 998:1996, Rakov and Uman 2003). The lower limit to the shaded area is near the curve recommended by Golde (1945, 1973) for negative lightning to flat Earth, for which there appears to be some experimental support, as outlined in the next paragraph. Because the basic physical parameters of the attachment process are not well known and, indeed, will vary from situation to situation, one should not expect any given expression like Eq. (3.2), with given values of A and b, to represent a reasonable approximation to any given situation. But, on average, there may be an expression that is generally reasonable, and one needs to adopt such an expression to allow the design of lightning protection systems to proceed. The degree of success of such an approach serves as some validation of the approach. Additional comments on Eq. (3.2) are found in Section 14.2. Most of the proposed expressions found in the literature relating peak current to striking distance were developed as part of "electrogeometric models" intended for application in electric power-line protection studies (see Chapter 12) where such models have had considerable practical success. For example, for strikes to the overhead wires of power lines, Armstrong and Whitehead (1968) suggest using $A = 6.7$ and $b = 0.8$ in Eq. (3.2), an increase of about 10 percent in the striking distance over the value they give for flat Earth for the same peak current (curve d_1 in Fig. 3.6). Some formulations of the relation between striking distance and peak current to overhead power lines are more complex than this: they develop expressions similar to Eq. (3.2) but containing some function that takes explicit account of the height of the lines (e.g., Eriksson 1987).

The length of the upward-connecting leader can sometimes be reasonably (questionably?) estimated from still photographs of lightning since the lightning channel sometimes changes direction sharply after the last step of the downward stepped leader and/or where the upward and downward leaders have met (see, for example, Fig. 1.8), although Berger and Vogelsanger (1966) show a streak photograph in their Figure 12 for which such features would not be obvious on a still photograph. Occasionally, multiple split channels, forming what looks like a loop or loops, can be seen at the junction point of upward and downward leaders, an example being given in Fig. 3.7. In long laboratory spark experiments similar loops are observed at the connection of upward and downward leaders, the basis for an extrapolation to the case of lightning. Sometimes the striking distance is considered to be twice the length of the upward leader, or twice the height to the channel loops, although there is not much evidence for this supposition. From the limited data available, photographically observed striking distances have been estimated to be between about 10 m and a few hundred meters (e.g., Golde 1977, Uman 1987, Rakov and Uman 2003). The shortest is perhaps the photograph described by Hagenguth (1947) of a lightning strike to a patch of weeds in a lake. The channel splits about 3 m above the water and remains split to a height of about 9 m, perhaps an upward leader length of about 6 m. Golde (1973) shows a photograph exhibiting a similar channel split (reproduced in Fig. 3.7), from which he infers that the meeting of the upward and downward leaders occurred 9 m above the chimney of the small building. Orville (1968) photographed a lightning strike to a 7-m-tall European ash tree, reproduced in Fig. 11.3, in which the upward-connecting leader extends 12 m above the tree top, as determined from the observation that upward-directed branches associated with the upward leader are present in the lower part of the lightning channel and downward-directed branches associated with the downward stepped leader are apparent in the upper part. Eight values of striking distance to a 60 m tower located on flat ground in South Africa (the tower on which the current records of Fig. 1.9 were obtained) were estimated from still photos to be between about 60 and 300 m for peak currents between about 10 and 100 kA (Eriksson 1978). The data show a general increase in striking distance for increased peak current and are located in the upper half of and just above the shaded area in Fig. 3.6. Striking distances to towers are expected to be somewhat greater than the calculated striking distances to flat ground illustrated in Fig. 3.6 because of the electric field enhancement at the tower top. Note again, however, that striking distance estimates from still photographs are subject to considerable error. Striking distances are best estimated from streak photographs such as that in Fig. 1.8, but even with these records it is not an unambiguous measurement. Golde (1973) sketches a streak photograph from South Africa showing an upward-leader length of about 50 m above flat ground. Berger and Vogelsanger (1966) in their Figure 12 show a streak photograph of lightning striking the 55 m of tower on Mount San Salvatore from which they measure a striking distance of about 27 m for a peak return stroke current of

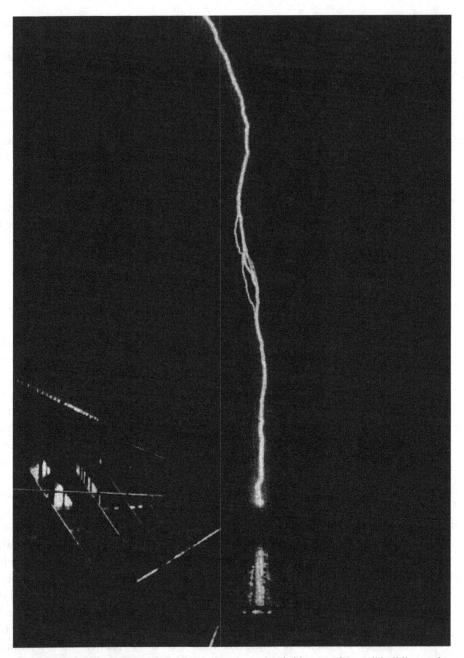

Fig. 3.7   A still photograph of a lightning strike to the unprotected chimney of a small building at the University of Helsinki. The loops in the lightning channel show the location where the upward and downward leaders connect. Taken from Golde (1973). Reproduced by permission of Edward Arnold (Publishers) Ltd.

16 kA; and in their Figure 13, reproduced in Fig. 1.8, a striking distance of about 37 m (straight-line from last leader step to tower) for a peak return stroke current of 27 kA. Note that in this latter case the last leader step was about 30 m above the tower and 30 m to its side when the upward leader was initiated, and that the connection was far from a straight line. Further, in both cases presented by Berger and Vogelsanger (1966), it appears that the upward leader represents a relatively large portion of the striking distance (see discussion in Section 1.3). The striking distance vs. peak current associated with the two streak photographs of Berger and Vogelsanger (1966) falls at the bottom of the shaded area in Fig. 3.6.

3.4 Cone of protection and rolling sphere methods

In the simplest version of the electrogeometric model, one that has been applied to power-line protection studies, the striking distances from the leader tip to the phase (high voltage) wires of a power line, to the overhead ground wires of the line, and to the Earth are all assumed to be equal. Then, for an assumed peak return stroke current, there is a unique striking distance via an expression like Eq. (3.2); and the lightning is predicted to terminate on whichever conducting object (various wires, the Earth) the tip of the leader first approaches to within that unique striking distance. The assumption of equal striking distances to each wire of a power line and to Earth ignores the fact that there are the different levels of electric field at the different potential strike-points because the shape of the conductor influences the magnitude of the electric field. Thus, the striking distances to the overhead ground wires, phase conductors, and Earth should not, in general, be equal and are not considered so in the more sophisticated electrogeometric models, although they generally are assumed to differ by only 10 percent or so. While electrogeometric models have been useful in many contexts, as in the design of the location of overhead ground wire protection on power lines (see Figs. 1.9, 12.2, and 12.3) and in the prediction of lightning outage rates on existing and newly designed power lines via Monte Carlo calculations (see Section 12.2.1), it is of value to have simpler approaches. Two such approaches, which also allow one to specify the placement of lightning rods or other overhead conductors to intercept the downward-moving stepped leader, are termed the "cone of protection" method (illustrated in Figs. 3.8 and 10.2) and the "rolling sphere" method (illustrated in Fig. 3.9a and b, and Fig. 12.6). The cone of protection method has had a history of over 200 years, well preceding the electrogeometric models, although those models can be used to justify it. The concept of a "rolling sphere" is relatively recent and is directly related to the electrogeometric models in that it relies directly on the assumption that a given striking distance is associated with a unique first-stroke peak current.

The determination of the region in which lightning cannot strike, that is, the region that is protected against lightning because it strikes somewhere else prefer-entially, has been a subject of discussion since the time of Benjamin Franklin.

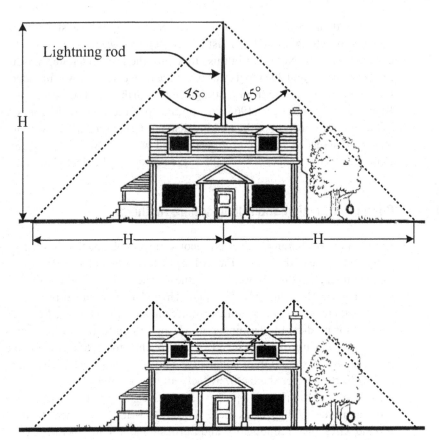

Fig. 3.8   The cone of protection method assuming a 45° cone with a single lightning rod (top), and multiple rods (bottom). Note that there are unprotected parts of the tree and, in the lower figure, of the roof that would, in theory, be protected if a 60° cone of protection had been assumed. Adapted from Uman (1986).

In 1777 a building in Purfleet, England, near London, that was used to store explosive materials sustained lightning damage on the edge of the roof at a horizontal distance of about 38 feet (11.6 m) from a vertical lightning rod mounted at the center of the structure. The top of the rod was 24 feet (7.3 m) above the roof edge, the lowest part of the roof. The ratio of the maximum horizontal distance from the rod within which the lightning was thought to be unable to strike (in this case, about 38 feet) to the rod height (24 feet) was termed the protective ratio (see Section 1.5), which for this example was equal to $(38/24) \cong 1.6{:}1$. The lightning protection system on this structure had been designed by a committee of which Benjamin Franklin was a member (Cavendish *et al.* 1773). A sketch of the building and other information is found in Golde (1977). Apparently, this was the first recorded case of the limited protection (the building was struck) provided by a grounded lightning rod, although the damage was minor considering that the building housed explosives. About 50 years later, in what might be considered the

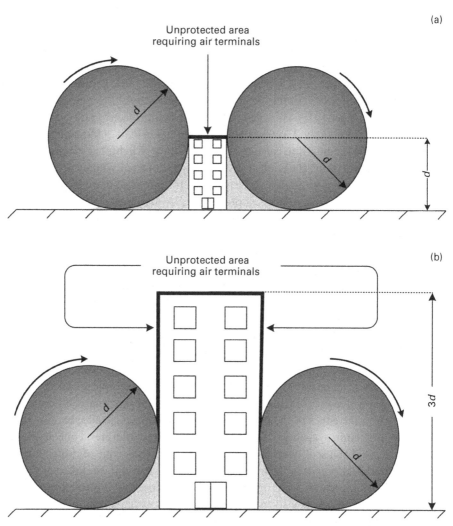

Fig. 3.9 Two examples of the application of the rolling sphere method. (a) The sphere is rolled over a structure whose height is equal to or less than the sphere radius (striking distance). The structure can only be struck by lightning on its top. (b) The sphere is rolled over a structure whose height is greater than the sphere radius (striking distance). The structure can be struck by lightning on its sides as well as its top. In both (a) and (b) the shaded volume is protected whereas the darkened surfaces on the structures may be struck.

earliest standard for lightning protection (Gay-Lussac and Pouillet 1823), the French Academy of Sciences concluded that a vertical rod would protect a circular area around its base whose radius was twice the rod height, a protective ratio of 2:1. Apparently, the Purfleet case was not considered or was forgotten. Additionally significant, from an historical point of view, are the laboratory work of Preece (1880) who concluded that the proper protective ratio was 1:1, the book by

Anderson (1879) who summarized all previous literature on lightning protection, the 1882 Report of the Lightning Rod Conference (Symons 1882) that established the standards for lightning protection in Britain, and the first American standard, issued in 1904 by the National Fire Protection Association (NFPA) (Lemmon et al. 1904), which was essentially similar to the 1882 British recommendations. In publications over time, proposed protective ratios have ranged between 0.1 and 9.0 (Schwaiger 1938). As noted in Section 1.5, a protective ratio of 3:1 was adopted by the International Electrotechnical Commission in its 2006 standard and by NFPA in its 2004 standard to calculate the probability of lightning striking a small building.

Historically, the zone of protection or protected volume provided by a lightning rod or other grounded vertical conductor has most often been considered to be an imaginary cone, a "cone of protection," whose apex is at the top of the conductor. Illustrations of the use of the cone of protection are found in Fig. 3.8 and Fig. 10.3. The concept of a cone of protection is adequate for the design of protection for small structures, boats, and trees; that is, even though there may be an occasional failure for a given cone angle, for reasons to be discussed later, protection design based on the cone of protection is not much different from that derived using other techniques. For a protective ratio of 1:1, the angle between the rod and the lateral surface of the cone at its apex is 45°, and the protected zone is usually referred to as a 45° cone. A protective ratio of 2:1 corresponds to approximately a 60° cone. Until recently, 45° and 60° cones of protection were commonly recommended in lightning protection standards. This is still the case in NFPA 780:2004 for structures with heights less than 15 m (50 feet). In the example given in Section 1.5, the protective ratio is assumed to be 2:1, or an approximate 60° cone of protection, from the edges of the roof. The larger the cone angle adopted, the more likely the protection system will fail. For very sensitive structures, cone angles less than 45° should be used. In specifying a cone angle for the cone of protection method prior to the latter part of the twentieth century, no account was taken of the physics of the lightning leader or the attachment process since that information was not available. Nevertheless, as we shall see, recent attempts have been made to relate the cone angle to the physical behavior of lightning.

The "rolling sphere" method to determine the placement of lightning rods or other overhead intercepting conductors is the simplest practical application of the electrogeometric model. The rolling sphere method is generally credited to Lee (1978, 1979). In that method, the tip of the leader is considered to be located at the center of an imaginary sphere whose radius is the striking distance. If such an imaginary sphere of a given radius, corresponding to a given peak current (related via Eq. [3.2]), is rolled along the Earth and over objects on the Earth, every point which is touched by the sphere is a possible point of strike, whereas points not touched are not, as illustrated in Fig. 3.9a and b. Thus, the sphere should touch lightning rods, vertical conducting masts, and other conductors intended to intercept the lightning but should not touch any part of the structure to be protected. Since the radius of the rolling sphere decreases with a decrease in the first return stroke peak current, there will always be a small value of peak current below which the system protection will fail because a sphere of small enough radius will roll between the lightning-intercepting conductors. On the other hand, there may be a minimum value for the first stroke peak current, below which the charge on the stepped leader is too low for the leader to propagate all the way downward to the Earth. Or, at least, there are not many return strokes with very small currents, and protection failure for such small currents may not have serious consequences. According to IEC 62305-1,3:2006, complete protection against 99 percent of all ground flashes can be accomplished by using a rolling sphere radius of 20 m (about 66 feet), and complete protection against 84 percent of all ground flashes by using a radius equal to 60 m (Section 4.2, Table 4.1). Since the relation between the rolling sphere radius (striking distance) and the first stroke peak current comes from adopting an expression like Eq. (3.2), which may not be accurate in any given case, the protection should not always be expected to work as designed. NFPA 780:2004 recommends using a rolling sphere radius (striking distance) of either 30 m (about 100 feet) which theoretically protects completely against 97 percent of all ground flashes (see Section 4.2, Table 4.1), or 46 m (about 150 feet) which theoretically protects completely against 91 percent of all ground flashes, for the design of the lightning protection of various structures, boats, and trees. It is interesting to note that if a wire mesh is laid directly on the roof of a structure so that it acts as the air terminal, as allowed by IEC 62305-1,3:2006 and some national standards, the rolling sphere method will predict that lightning can strike the roof inside the mesh, but experience apparently does not indicate that this failure occurs.

The rolling sphere approach predicts that tall structures can be struck on their sides above a certain height, as illustrated in Fig. 3.9b. This is not common for structures less than about 60 m in height but does occur for structures over about 120 m, according to observation and to the rolling sphere model (e.g., IEC 62305-3:2006). Figure 3.10 shows an example of this situation in which a downward leader moving at an angle relative to the vertical has not come close enough to the top of Toronto's CN tower to elicit an upward leader from the tower top but does apparently get within its striking distance of the building's side at a distance of 45 m below the top. From 1991 to 2005 there were 404 lightning attachments to the CN tower of which 16 (4 percent) were below its tip, the distances below the tip being 5.4 to 70 m (Hussein *et al.* 2007). If the striking distance is the horizontal portion of the channel in Fig. 3.10, its length is about 50 m. Note that for the 553-m-tall CN tower most of the lightning is upward-initiated (see Sections 1.1 and 1.2), while one might expect that it would be mostly downward-initiated lightning that would strike below the tower top.

For a lightning-protected structure, the protected volume determined by rolling a sphere of any given radius over the structure can be used to define a cone angle for the cone of protection, as discussed in more detail in Section 4.2. For small structures, those with heights of about 10 to 20 m or less, using a 45° cone (a protective ratio of 1:1), a 60° cone (a protective ratio of about 2:1), or a 70° cone (a protective ratio of about 3:1) in the cone of protection method is more or less

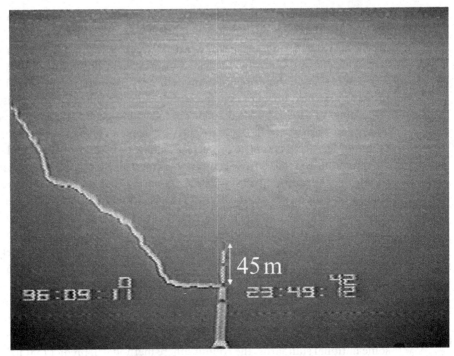

45 m

Fig. 3.10   A video frame showing lightning striking 45 m below the top of the 553-m-high CN Tower
in Toronto, Canada. Courtesy of Ali Hussein and Wasyl Janischewskyj.

equivalent to the rolling sphere method for a reasonable range of sphere radii and for
building heights smaller than the striking distance (compare Fig. 3.8 with Fig. 3.9a).

Theoretical protection techniques more complex than the single rolling sphere
approach, in that they take account of the fact that different system elements have
different striking distances because of the different electric field enhancement at
those elements or of additional issues of physics or geometry, are known as the
"geometric zone of capture," the "collection volume method," and the "collection
surface method," but they do not add sufficiently to our discussion to consider
them here.

References

Anderson, R. 1879. *Lightning Conductors: – Their History, Nature, and Mode of Application*.
London: E. and F. N. Spon.

Anderson, J. G. and Tangen, K. O. 1968. Insulation for switching-surge voltages, in *EHV
Transmission Line Reference Book*. New York: Edison Electric Institute. pp. 215–256.

Armstrong, H. R. and Whitehead, E. R. 1968. Field and analytical studies of transmission
line shielding – III. *IEEE Trans. Pow. Appar. Syst.* **87**: 270–281.

Bazelyan, E. M. and Raizer, Y. 1997. *Spark Discharge*. Boca Raton: CRC Press.

Bazelyan, E. M. and Raizer, Y. 2000. *Lightning Physics and Lightning Protection*. Bristol and
Philadelphia: Institute of Physics Publishing.

Berger, K. 1972. Methoden und Resultate der Blitzforschung auf dem Monte San Salvatore bei Lugano in den Jahren 1963–1971. *Bull. Schweiz. Elektrotech. Ver.* **63**: 1403–1422.

Berger, K. and Vogelsanger, E. 1966. Photographische Blitzuntersuchugen der Jahre 1955 … 1965 auf dem Monte San Salvatore. *Bull. Schweiz. Elektrotech. Ver.* **57**: 599–620.

Cavendish, H., Watson, W., Franklin, B. and Robertson, J. 1773. Report of the committee appointed by the Royal Society to consider a method of securing the powder magazine at Purfleet. *Phil. Trans. Roy. Soc.* **63**: 42–47.

Chowdhuri, P. 1996. *Electromagnetic Transients in Power Systems.* New York: John Wiley and Sons.

Cooray, V., Rakov, V. and Theethayi, N. 2007. The lightning striking distance – revisited. *J. Electrostat.* **65**: 296–306.

Dwyer, J. R., Rassoul, H. K., Al-Dayeh, M. *et al.* 2005. X-ray bursts associated with leader steps in cloud-to-ground lightning. *Geophys. Res. Lett.* **32**: L01803, doi:10.1029/2004GL021782, 2005.

Eriksson, A. J. 1978. *A Discussion on Lightning and Tall Structures, CSIR Special Report ELEK 152,* National Electrical Engineering Research Institute, Pretoria, South Africa, July 1978. See also Lightning and tall structures. *Trans. S. Afr. Electr. Eng.* **69**: 238–252.

Eriksson, A. J. 1987. An improved electrogeometric model for transmission line shielding analysis. *IEEE Trans. Pow. Del.* **2**: 871–886.

Franklin, B. 1767, Letter XXIV, in Cohen I. B. 1941. *Benjamin Franklin's Experiments.* Cambridge, MA: Harvard University Press, pp. 388–392.

Gay-Lussac, F. and Pouillet, C. 1823. *Introduction sur les paratonnères, adoptée par l' Académie des Sciences.* Paris.

Golde, R. H. 1945. On the frequency of occurrence and the distribution of lightning flashes to transmission lines. *AIEE Trans.* **64**(III): 902–910.

Golde, R. H. 1973. *Lightning Protection.* London: Edward Arnold (Publishers) Ltd.

Golde, R. H. 1977. The lightning conductor. In *Lightning,* Vol. II: *Lightning Protection,* ed. R. H. Golde. New York: Academic Press, p. 545–576.

Hagenguth, J. H. 1947. Photographic study of lightning. *Trans. Am. Inst. Electr. Eng.* **66**: 577–585.

Hussein, A. M., Milewski, M., Janischewskyj, W., Noor, F. and Jabbar, F., 2007. Characteristics of lightning striking the CN tower below its tip. *J. Electrostat,* **65**: 307–315.

IEC 62305-1:2006. *Protection Against Lightning. Part 1: General Principles.* Geneva: International Electrotechnical Commission.

IEC 62305-2:2006. *Protection Against Lightning. Part 2: Risk Management.* Geneva: International Electrotechnical Commission.

IEC 62305-3:2006. *Protection Against Lightning. Part 3: Physical Damage to Structures and Life Hazard.* Geneva: International Electrotechnical Commission.

IEC 62305-4:2006. *Protection Against Lightning. Part 4: Electrical and Electronic Systems within Structures.* Geneva: International Electrotechnical Commission.

IEC 62305-5:2006. *Protection Against Lightning. Part 5: Services* (to be published). Geneva: International Electrotechnical Commission.

IEEE Standard 998:1996. *IEEE Guide for Direct Lightning Stroke Shielding of Substations.* Geneva: Institute of Electrical and Electronics Engineers, Inc.

Lee, R. H. 1978. Protection zone for buildings against lightning strokes using transmission line protection practice. *IEEE Trans. Ind. Appl.* **14**: 465–470.

Lee, R. H. 1979. Lightning protection of buildings. *IEEE Trans. Ind. Appl.* **15**: 236–240.

Lemmon, W. S., Loomis, B. H. and Barbour, R. P. 1904. *Specifications for Protection of Buildings Against Lightning*. Quincy, MA: National Fire Protection Association.

Les Renardières Group. 1977. Positive discharges in long air gaps at Les Renardières, 1975 results and conclusions. *Electra* **53**: 31–153.

Les Renardières Group. 1981. Negative discharges in long air gaps at Les Renardières, 1978 results. *Electra* **74**: 67–216.

NFPA 780:2004. *Standard for the Installation of Lightning Protection Systems*. Quincy, MA: National Fire Protection Association (NFPA).

Orville, R. E. 1968. Photograph of a close lightning flash. *Science* **162**: 666–667.

Preece, W. H. 1880. On the space protected by a lightning conductor. *Phil. Magazine* **9**, 427–430.

Rakov, V. A., and Uman, M. A. 2003. *Lightning: Physics and Effects*. Cambridge: Cambridge University Press.

Rizk, F. A. M. 1990. Modeling of transmission line exposure to direct lightning strokes. *IEEE Trans. Power Delivery* **5**, No. 4, 1982–1997.

Rizk, F. A. M. 1994a. Modeling of lightning incidence to tall structures, Part I: Theory. *IEEE Trans. Power Delivery* **9**, No. 1, 162–171.

Rizk, F. A. M. 1994b. Modeling of lightning incidence to tall structures, Part II: Application. *IEEE Trans. Power Delivery* **9**, No. 1, 172–193.

Sadiku, M. 2007. *Element of Electromagnetics*, 4th edn. Oxford: Oxford University Press

Schwaiger, A. 1938. *Der Schutzbereich von Blitzableitern*. Munich: R. Oldenbourg.

Symons, G. J., ed. 1882. Report of the Lightning Rod Conference (with delegates from the following societies, viz: Meteorological Society, Royal Institute of British Architects, Society of Telegraph and of Electricians, Physical Society, Co-opted members [Prof. W. E. Ayrton, Prof. D. E. Hughes]). London: E. & F. N. Spon.

Tesche, F. M. 1978. Topological concepts for internal EMP interaction. *IEEE Trans. Electromagn. Compat.* **20**: 60–64.

Uman, M. A. 1986. *All About Lightning*. New York: Dover Publications, Inc. (Revised paperback edition of *Understanding Lightning*).

Uman, M. A. 1987. *The Lightning Discharge*. London: Academic Press. (Revised paperback edition, 2001. New York: Dover Publications, Inc.).

Vance, E. F. 1980. Electromagnetic interference control. *IEEE Trans. Electromagn. Compat.* **22**: 319–328.

4 Structure protection: air terminals and down conductors

4.1 Overview

More than 250 years ago Benjamin Franklin (1753) described the basis of a scheme for the protection of structures:

How to secure Houses, etc. from Lightning
It has pleased God in his Goodness to Mankind, at length to discover to them the Means of securing their Habitations and other Buildings from Mischief by Thunder and Lightning. The Method is this: Provide a small Iron Rod (it may be made of the Rod-iron used by the Nailers) but of such a Length, that one End being three or four Feet in the moist Ground, the other may be six or eight Feet above the highest Part of the Building. To the upper End of the Rod fasten about a Foot of Brass Wire, the Size of a common Knitting-needle, sharpened to a fine Point; the Rod may be secured to the House by a few small Staples. If the House or Barn be long, there may be a Rod and Point at each End, and a middling Wire along the Ridge from one to the other. A House thus furnished will not be damaged by Lightning, it being attracted by the Points, and passing thro the Metal into the Ground without hurting any Thing. Vessels also, having a sharp pointed Rod fix'd on the Top of their Masts, with a Wire from the Foot of the Rod reaching down, round one of the Shrouds, to the Water, will not be hurt by Lightning.

Although the general principles are clear and were elucidated by Franklin, the detailed design of lightning protection systems, including the optimum geometry of lightning rods (i.e., the length, the diameter, the curvature of the tip) and the necessary number and location of the rods has been argued since the time of Franklin. It is primarily 250 years of experience that validates modern structure protection such as specified in the various standards, although, as we have seen in Section 3.3 and Section 3.4, there is some level of supporting theory and a continuing effort to improve that theory.

Examples of lightning protection systems that are attached directly to the structures to be protected and that involve lightning rods were illustrated in Figs. 3.3–3.5. Similarly effective lightning protection systems can also be isolated from a protected structure, examples being (1) the vertical insulating masts with grounded wires on top that protect some space vehicles at the NASA Kennedy Space Center and associated US Air Force launch facilities, such as the Atlas V (Fig. 4.1), the Space Shuttle (Fig. 4.2), and, previously, the Apollo Series of Saturn V moon rockets, (2) the tall conducting masts that often surround oil storage tanks, with or without wires connecting the mast tops, and (3) the overhead ground wires

Fig. 4.1   Atlas V launch vehicle lightning protection. Insulating masts on four towers support overhead wires (the wire diameters are computer enlarged for clarity) connecting the mast tops and extending down to grounding electrodes some distance away. The vehicle shown was launched in early 2006 and carries a probe intended for the "dwarf" planet Pluto. Courtesy of NASA.

Fig. 4.2   Space Shuttle lightning protection. A computer-generated photograph of the Space Shuttle and its overhead-wire (catenary wire) lightning protection system. The catenary wire is about 300 m long in either direction. Its diameter has been computer-enhanced for clarity. Courtesy of NASA.

on some power lines (Fig. 1.10, Fig. 12.2, and Fig. 12.3). A successful operation of the shuttle protection system is illustrated in Fig. 4.3. In general, isolated protection systems are used for facilities that require a very high level of protection, such as a fuel or explosives storage areas or rocket-launch facilities. In these cases the close

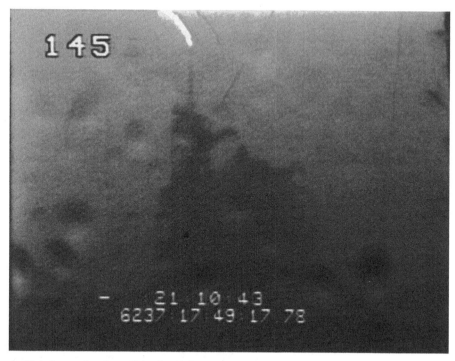

Fig. 4.3   A video frame showing a lightning strike to the highest point of the Space Shuttle protection system. The Shuttle Atlantis was on the pad, ready for launch, on August 25, 2006, when the lightning strike occurred. Courtesy of NASA.

lightning channel itself or lightning-induced sparking from the close lightning channel generally cannot be tolerated because of the potential combustion of flammable or explosive gas vapors or explosive solid materials, or the potential induction of deleterious voltages into sensitive electronic systems from the electric and magnetic fields of the close lightning.

4.2   Air terminals

Nearly everyone has seen lightning rods mounted on the roofs of buildings; probably every individual who has enough interest to read this book. A photograph of a combination lightning rod, down conductor, and grounding electrode dating from the late 1700s is shown in Fig. 4.4. Examples of modern lightning rods and more elaborate and ornate older versions are shown in Fig. 1.11 and Figs. 4.5–4.9. All elevated metal structures intended to intercept the lightning are generically known as "air terminals," lightning rods being the most common example. There is no evidence that the many fancy lightning rod accoutrements such as glass balls, horizontal arrows, and multipronged tips sometimes seen on older houses and barns serve any scientific purpose. A wide variety of such beautiful "rods," an

Fig. 4.4   An eighteenth-century house outfitted with a lightning rod of Benjamin Franklin's 1762 design. The rod continues downward as down conductor and grounding electrode. Courtesy of E. Philip Krider.

example being shown in Fig. 4.9, were manufactured and marketed, particularly in the nineteenth century in the US Midwest. These are now found in antique stores at prices that can be $1000 or more. Rare glass balls can cost up to $10 000.

As noted in Section 3.2, the term "Franklin rod" is often used to describe any lightning rod, but sometimes the term is specifically meant to identify rods with

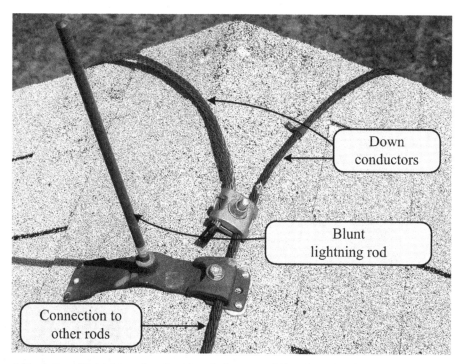

Fig. 4.5 A blunt-tipped lightning rod on the edge of a house roof. Two down conductors are attached, each leading to a grounding electrode on a different corner of the house. NFPA 780:2004 states that the lightning rod should be located within 0.6 m (2 feet) of the edge of the roof. Photograph by Brian De Carlo.

sharp tips, the geometry originally recommended by Franklin, as indicated in the quote in Section 4.1. Recently, experimental evidence has been published that moderately blunt rod tips are more effective than sharp tips (Moore *et al.* 2000a,b). In field experiments where pairs of sharp and blunt rods of equal height separated by 5 to 20 m were erected on a mountain top in New Mexico, it was found over a 5-year period that no sharp rods were struck by lightning. However, 12 blunt rods with diameters from 1.27 to 2.54 cm (1/2 to 1 inch) were struck. Apparently, five to eight pairs of rods were used each year. Moore *et al.* (2000a,b) conclude that the strike probability for lightning rods is increased when their tips are made moderately blunt (ratio of tip height to tip radius of curvature of about 680:1) as opposed to sharper rods or very blunt ones. It would certainly be valuable to have confirmation of these experimental results by other investigators at different locations. Such confirmation would likely lead to a standard for the radius of curvature of the rod tip. The physical argument for the ineffectiveness of sharp rods is that sharp rods produce corona discharge in lower values of the ambient electric field than do blunt rods (because sharp rods provide greater enhancement of the ambient electric field) and the corona discharge inhibits the launching of an upward-connecting leader. NFPA 780:2004 states that either blunt or sharp tips may be

Fig. 4.6   A moderately blunt-tipped rod on the roof of a bus shelter at the University of Florida. Photograph by Keith Rambo.

used, without stating that one is better than the other, but requires that any rod used must exceed 10 inches (25.4 cm) in height. The trend among lightning rod installers is to use rods with blunt tips, probably in view of the research noted above.

Ideally, any air terminal should be designed to withstand lightning without significant damage to itself, although such damage is not necessarily harmful. Conducting spires and roof edges, metal roofs, and various vertical metal components above the roof level can be used as air terminals if potential melting or burn-through of these metallic objects is not a problem for other reasons. Several such reasons are noted later.

An example of an unusual air terminal was the aluminum pyramid set on top of the newly built Washington Monument in 1884, then the tallest man-made structure in the world. At the time, the pyramid was the largest aluminum casting ever made, termed "the crown jewel of the aluminum industry," and as such it was displayed at Tiffany's in New York before its installation on the Monument (Dix 1934, Binczewski 1995). The pyramid was removed in 1934 when the Monument

Fig. 4.7 A sharp-tipped rod on the roof of Benton Hall at the University of Florida. Note the stranded cable bonding the rod to all other rods on the roof and to the down conductors. Photograph by Keith Rambo.

was renovated. The pyramid showed some lightning damage, as was expected, but was substantially intact (Dix 1934).

    IEC 62305-1,2,3,4,5:2006, NFPA 780:2004, and many other national standards recommend installing multiple connected lightning rods, but similar protection is apparently obtained by covering the insulating roof of a building with a mesh of metallic conductors. IEC 62305-1,2,3,4,5:2006 and many national codes allow the

Fig. 4.8   A sequence of sharp-tipped rods on the roof of the Engineering Building at the University of Florida. Note the stranded cable connecting all rods both together and to the down conductors. Photograph by Britt Hanley.

metal meshes. The usual size of the mesh recommended is between 5 and 20 m. The ultimate air terminal is a solid metal roof, provided it is of adequate thickness (NFPA 780:2004 specifies greater than 3/16 inch (4.8 mm); IEC 62305-3:2006 specifies 4 mm for steel, 5 mm for copper, and 7 mm for aluminum) to avoid lightning penetration at the point of strike if such penetration is a problem (1) because of fire below roof level that may potentially be caused by dripping hot metal or (2) because of water damage to the interior via rain passage through any lightning-burned hole in the roof. Additionally, metal roofs are sometimes covered with tar-like roofing material, and this potentially could be set on fire by molten roof metal which may remain hot after the lightning current has ceased.

Steel supports and well-bonded rebar in the reinforced concrete of walls and roofs may be used both for part of the air-terminal system and for down conductors (Fig. 3.5). Otherwise, separate air terminals should be well bonded to the metal supports and rebar at roof level.

Tall structures may be subjected to lightning strikes from the side, as has often been observed, and as is predicted by the rolling sphere method (Sections 3.3 and 3.4; Fig. 3.9b, and Fig. 3.10). Buildings exceeding 20 to 30 m (about 65 to 100 feet)

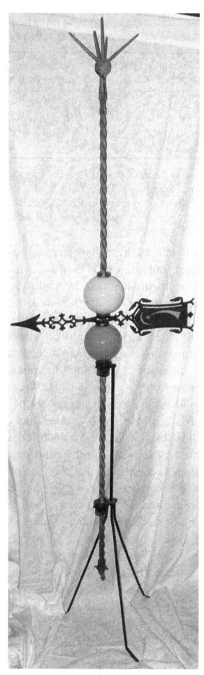

Fig. 4.9   A nineteenth-century rod that is about 2 m (over 6 feet) in height with a fancy tip (finial), ornamental glass balls, and an ornamental arrow with red glass insert. Photograph by Derek Uman.

Table 4.1 Protection levels defined in IEC 62305-1,3:2006 and corresponding rolling sphere radius, minimum peak current, and percentage of first strokes with peak currents greater than the minimum.

Protection level	Rolling sphere radius R, meters	Minimum peak current I, kA	Percentage of first strokes greater than minimum
I	20	3	99
II	30	5	97
III	45	10	91
IV	60	16	84

The author thanks the International Electrotechnical Commission (IEC) for permission to reproduce information from its International Standard IEC 62305-1 ed. 1.0 and IEC 62305-3 ed. 1.0. All such extracts are copyright of IEC, Geneva, Switzerland. All rights reserved.

in height should therefore be provided with additional air terminals on their side walls. Metallic facades, window-frames, railings, and exposed down conductors can be used for this purpose. Particular attention should be paid to surfaces at a height above 60 m since they are almost certain to be struck eventually.

For the past decade or so, the rolling sphere approach has been the method of choice for determining the placement of air terminals on structures. The International Electrotechnical Commission (IEC) has specified four levels of lightning protection, each related to the percentage of total lightning events for which complete protection is achieved. Level I protection, the highest level of protection, theoretically provides complete protection against 99 percent of all cloud-to-ground flashes, that is, all first stroke currents greater than or equal to 3 kA. The rolling sphere radii specified by IEC 62305-1,3:2006 for each of the four levels of protection, taking into account both positive and negative lightning, are given in Table 4.1. According to the theory, for a given rolling sphere radius, all flashes with first stroke peak currents higher than the peak current value associated with the given sphere radius (via Eq. [3.2]) will be intercepted by the air terminals. Smaller first stroke peak currents may or may not be intercepted, depending on the stroke location, so there is some degree of protection for these smaller strokes also. Note from Table 4.1 that for first stroke peak currents smaller than about 3 kA (estimated in Table 4.1 to be 1 percent of the population of first stroke currents) even the use of a 20 m radius (about 65 feet) for the rolling-sphere protection design will not be sufficient to assure complete protection. Protection system design using a 60 m sphere (about 200 feet) is predicted to protect completely against 84 percent of flashes. But, as indicated above, the remaining 16 percent of flashes can still strike the air terminals of the 60 m sphere if the trajectory of their stepped leaders brings them near enough to those terminals, so the system failure rate is considerably less than the 16 percent that is sometimes erroneously inferred. The relation between rolling sphere radius and peak current used by the IEC to derive the values in Table 4.1, Eq. (3.2) with $A = 10$, $b = 0.65$, is plotted as d_3 in Fig. 3.6. The other information

Table 4.2 Relation between rolling sphere radius and protection system mesh size for different protection levels adapted from IEC 62305-3:2006.

Protection level	Rolling sphere radius, meters	Mesh size, meters by meters
I	20	5×5
II	30	10×10
III	45	15×15
IV	60	20×20

The author thanks the International Electrotechnical Commission (IEC) for permission to reproduce information from its International Standard IEC 62305-1 ed. 1.0 and IEC 62305-3 ed. 1.0. All such extracts are copyright of IEC, Geneva, Switzerland. All rights reserved.

needed to complete Table 4.1 is the cumulative distribution of first stroke currents. This is obtained from extrapolation of the data of Berger *et al.* (1975) on 101 first stroke currents, the same data base from which Table 2.1 was developed. The rolling sphere radii given in Table 4.1 would not be changed too much if one of the other two curves plotted in Fig. 3.6 were used (in Eq. [3.2], $A = 6$, $b = 0.8$ or $A = 8$, $b = 0.65$) instead of the IEC's curve. Then, instead of the 60 m sphere radius for Protection Level IV found in Table 4.1, the calculated radius from Eq. (3.2) would be 54 m or 48 m, respectively. Instead of the 30 m sphere radius for Protection Level II, the calculated radius would be 23 m or 24 m, respectively. A more substantial change in sphere radius, a reduction of a factor of 3 to 4 from the IEC values, occurs if one adopts the lower limit to the shaded area in Fig. 3.6 as the proper relation between striking distance and peak current. In this case, the appropriate sphere radius for Protection Level IV would be about 20 m.

Table 4.2 gives the mesh sizes suggested by the IEC for the four protection levels of Table 4.1. As noted in Section 3.4, the rolling sphere method predicts that lightning can strike inside any mesh laid directly on a roof. No theory is presented by the IEC to justify the values of mesh size given.

The equivalent cone angle for the cone of protection method when applied to the side wall of a structure can be calculated for each rolling sphere protection level and any structure height. Figure 3.9a illustrates this equivalence. For Protection Level I, a 20 m rolling sphere radius, the equivalent cone angle is about 45° for a 10-m-high structure and about 20° for a 20-m-high structure. For a 10-m-high structure, the equivalent cone angle is about 55° for Protection Level II, about 60° for Level III, and about 65° for Level IV. For a 20-m-high structure, the Level II equivalent cone angle is about 40°, the Level III angle is about 50°, and the Level IV angle about 55°. It is clear that for small structures a cone angle of 45° to 60° is very reasonable and fully consistent with the rolling sphere approach. A cone angle of about 70°, the 3:1 protective ratio used by IEC and NFPA in risk calculations (see Sections 1.5 and 1.6), is consistent with a structure height of 2 to 6 m for Levels I to IV, respectively. All the numbers given above are taken from IEC 62305-3:2006.

4.3   Early streamer emission rods

As we have seen, the purpose of an air terminal is to launch an upward-moving leader to "capture" the downward-propagating stepped leader. The purpose of the other two components of the protection system, the down conductors and the grounding electrodes, is then to dispose of the lightning current as harmlessly as possible. So-called "early streamer emission rods" are special air terminals that are claimed to initiate upward leaders earlier and to produce longer upward leaders than is the case for standard lightning rods. If this claim were true, fewer early streamer emission rods would be needed in a protection system than standard rods. In fact, according to their advertising, generally only one early streamer emission rod is needed to protect a typical structure. The down conductors and grounding of a protection system using such rods are basically the same as for a standard system. Clearly, if the rods do not work as advertised, which is the present view of most members of the lightning community, the result is a potentially dangerous degree of underprotection. Most early streamer emission rods either use small amounts of radioactive material to ionize the air near the rod (such radioactive air terminals being specifically forbidden by IEC 62305-3:2006) or contain electronic circuitry and/or small spark gaps that provide small sparks for the same purpose, local air ionization, when the electric field at the rod reaches a prescribed level. An early streamer emission rod known as a Preventron is shown in Fig. 4.10. Others have the appearance of inverted salad bowls or flying saucers mounted on a pole.

We briefly discuss now the arguments for the ineffectiveness of early streamer emission rods. If an upward leader from an early streamer rod is to be initiated earlier than from a standard rod, it must be initiated in a smaller electric field than the latter since the electric field near the ground from the downward-moving stepped leader increases with time. It is not clear that an upward leader initiated in a lower field would propagate at the same speed as the leader from the conventional rod in a higher field. Once initiated, the early streamer upward leader might propagate more slowly or indeed not propagate at all. Proponents of early streamer emission devices claim (1) a $100\,\mu s$ initiation-time advantage over conventional rods and (2) upward-leader speeds of $1\,m\,\mu s^{-1}$. If true, the early streamer emission leader would travel 100 m further than the conventional leader in the extra $100\,\mu s$; hence the claimed increase in area coverage. Both aspects of the proponents' argument are questionable: (1) the few existing measurements of upward-positive leader speed during the attachment process indicate a typical speed near $0.1\,m\,\mu s^{-1}$ (Berger and Vogelsanger 1966, 1969, McEachron 1939, Yokoyama et al. 1990), ten times less than proponents assume, and (2) there are no data supporting the postulated $100\,\mu s$ advantage in initiation time. Even if the postulated time advantage were true and even if the early streamer emission leader could propagate effectively in a lower field, the typical measured leader speeds quoted above indicate only a 10 m advantage in collection range. Given that a typical upward

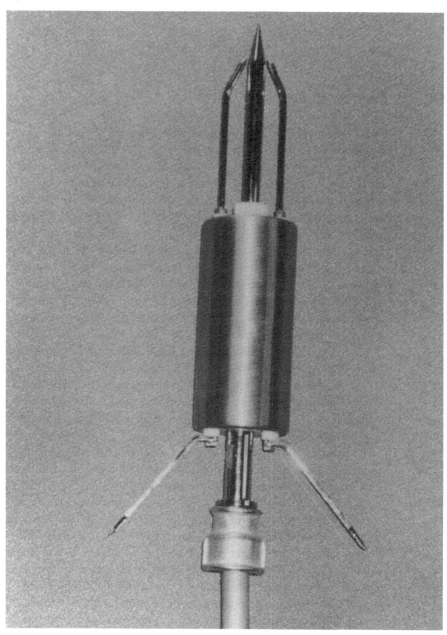

Fig. 4.10   A photograph of an "early streamer emission" air terminal.

positive leader has a length in the 50 m range (see Fig. 3.6), an extra 10 m is not much of an improvement. More details on early streamer emission protection systems are found in Uman and Rakov (2002).

4.4   Down conductors

The air terminals of a lightning protection system should be connected by the shortest possible route via the maximum possible number of down conductors to the grounding system. Two down conductors are shown, for example, in Fig. 4.5, and down conductors are illustrated by dashed lines in Figs. 3.3 and 3.4. A symmetrical arrangement of multiple down conductors (1) minimizes the inductive voltage drop in the overall down-conductor system, reducing voltage differences between objects or equipment connected to the down-conductor system at different heights and voltages that can shock individuals, and (2) minimizes the magnetic field inside the protected structure associated with the lightning current in the down conductors, the time-variation of that magnetic field being potentially responsible for damage to electronics (see Section 2.3, Fig. 2.7, and Fig. 2.8). Even a small structure should have a minimum of two down conductors, generally placed at opposite sides of the structure.

A specification by the International Electrotechnical Commission of materials for air terminals and down conductors is given in Table 4.3. According to IEC-62305-1,3:2006, the minimum cross-section of wire to prevent melting by the heating associated with very large action integrals (see Section 2.3) is 16 mm^2 for copper, 25 mm^2 for aluminum, 50 mm^2 for steel, and 100 mm^2 for stainless steel (see also footnote *i* in Table 4.3).

The recommended minimum cross-sectional areas for use in lightning protection system components found in Table 4.3 for copper and aluminum are about three times larger than the cross-sectional areas for which melting would occur for an action integral of $10 \times 10^6 \, \text{A}^2 \text{s}$. NFPA 780:2004 specifies a cross-sectional area of lightning conductor cable of 29 mm^2 for copper and 50 mm^2 for aluminum, for structures less than 23 m (75 feet) high. For taller structures, NFPA 780:2004 specifies 58 mm^2 for copper and 97 mm^2 for aluminum. According to Golde (1968), a No. 4 AWG (American Wire Gage) copper wire (cross-sectional area about 21 mm^2) will have a temperature rise of about 100 °C (180 °F) for an action integral of $5 \times 10^6 \, \text{A}^2 \text{s}$ and will melt at an action integral of about $20 \times 10^6 \, \text{A}^2 \text{s}$. A No. 8 AWG copper wire (about 8.4 mm^2) will melt at an action integral of about $5 \times 10^6 \, \text{A}^2 \text{s}$. For copper with the recommended minimum cross-sectional area of 50 mm^2, the temperature rise is 22 °C (40 °F) for an action integral of $10 \times 10^6 \, \text{A}^2 \text{s}$. Berger *et al.* (1975) report that the upper 5 percent of negative first strokes exhibit an action integral above $0.55 \times 10^6 \, \text{A}^2 \text{s}$, and the upper 5 percent of positive strokes exhibit an action integral above $15 \times 10^6 \, \text{A}^2 \text{s}$ (see Table 2.1). Maximum flash action integrals might be expected to be near $1 \times 10^6 \, \text{A}^2 \text{s}$ for negative flashes and near $50 \times 10^6 \, \text{A}^2 \text{s}$ for positive flashes. Since the probability of a structure's being

Table 4.3 Material, configuration, and minimum cross-sectional area of air-termination conductors, air-termination rods, and down conductors.

Material	Configuration	Minimum cross-sectional area	Comments[j]
Copper	solid tape	50 mm^{2h}	2 mm min. thickness
	solid round[g]	50 mm^{2h}	8 mm diameter
	stranded	50 mm^{2h}	1.7 mm min. diameter of each strand
	solid round[c,d]	200 mm^{2h}	16 mm diameter
Tin plated copper[a]	solid tape	50 mm^{2h}	2 mm min. thickness
	solid round[g]	50 mm^{2h}	8 mm diameter
	stranded	50 mm^{2h}	1.7 mm min. diameter of each strand
Aluminum	solid tape	70 mm^2	3 mm min. thickness
	solid round	50 mm^{2h}	8 mm diameter
	stranded	50 mm^{2h}	1.7 mm min. diameter of each strand
Aluminum alloy	solid tape	50 mm^2	2.5 mm min. thickness
	solid round	50 mm^2	8 mm diameter
	stranded	50 mm^2	1.7 mm min. diameter of each strand
	solid round[c]	200 mm^2	16 mm diameter
Hot dip galvanized steel[b]	solid tape	50 mm^{2h}	2.5 mm min. thickness
	solid round[i]	50 mm^2	8 mm diameter
	stranded	50 mm^{2h}	1.7 mm min. diameter of each strand
	solid round[c,d,i]	200 mm^{2h}	16 mm diameter
Stainless steel[e]	solid tape[f]	60 mm^{2h}	2 mm min. thickness
	solid round[f]	50 mm^2	8 mm diameter
	stranded	70 mm^{2h}	1.7 mm min. diameter of each strand
	solid round[c,d]	200 mm^{2h}	16 mm diameter

[a] Hot dipped or electroplated minimum thickness coating of 2 micrometers (microns).

[b] The coating should be smooth, continuous and free from flux stains with a minimum thickness coating of 50 micrometers.

[c] Applicable for air termination rods only.

[d] Applicable for Earth lead-in rods only.

[e] Chromium 16 percent, nickel 8 percent, carbon 0.1 percent max.

[f] For stainless steel embedded in concrete and/or in direct contact with flammable material the minimum sizes should be increased to 75 mm^2 (10 mm diameter) for solid round and 75 mm^2 (3 mm minimum thickness) for solid tape.

[g] 50 mm^2 (8 mm diameter) may be reduced to 28 mm^2 (6 mm diameter) in certain applications where mechanical strength is not an essential requirement. Consideration should, in this case, be given to reducing the spacing of the fasteners.

[h] If thermal and mechanical considerations are important, these dimensions can be increased to 60 mm^2 for solid tape and to 78 mm^2 for solid round.

[i] The minimum cross-section to avoid melting is 16 mm^2 (copper), 25 mm^2 (aluminum), 50 mm^2 (steel) and 50 mm^2 (stainless steel) for a specific energy of 10 000 kJ Ω^{-1}. For further information see Annex E.

[j] Thickness, width and diameter are defined at $\pm 10\%$.

The author thanks the International Electrotechnical Commission (IEC) for permission to reproduce information from its International Standard IEC 62305-1 ed. 1.0 and IEC 62305-3 ed. 1.0. All such extracts are copyright of IEC, Geneva, Switzerland. All rights reserved.

After IEC 62305-3:2006, Table 6.

struck by a very large action integral is exceedingly low, use of relatively small cross-section wire in a lightning protection system will work satisfactorily most of the time, although such wire is in violation of the various standards.

Almost-closed loops in down conductors are best avoided as the lightning might arc the shortest distance between wires in the loop (potentially through flammable material) owing to the voltage associated with the loop inductance. Hence, down conductors should generally be passed through large overhanging projections on buildings instead of being bent around them. Large metallic bodies within about 5 m of down conductors should be bonded to the down conductors in order to reduce the possibility of "side flashes." A justification of this maximum side flash length follows.

Consider a point on a vertical down conductor at a height h above ground. The situation is illustrated in Fig. 4.11. The bottom of the down conductor is grounded in the Earth with a resistance R_{gr}. The down-conductor wire has an inductance per unit length L_h. If current $I(t)$ flows in the down conductor, the voltage at the height h on the down conductor with respect to distant ground is the sum of the resistive voltage associated with the grounding connection and the inductive voltage associated with height h on the down-conductor wire.

$$V(t) = R_{gr}I(t) + L_h h \frac{\mathrm{d}I(t)}{\mathrm{d}t} \tag{4.1}$$

The resistive voltage component (the first term on the right of Eq. [4.1]) is present everywhere on the down conductor with the same value, $R_{gr}I(t)$, while the inductive component (the second term) increases linearly with height on the down conductor and is zero at the ground junction, $h=0$. A potential difference $V(t)$ will exist between a point on the protection system at height h and any separately grounded piece of metal (or a human standing on the ground). If $V(t)$ exceeds the breakdown voltage between the two conductors, there will be an electrical discharge, a "side flash," between them. We can roughly determine the potential maximum distance over which a side flash can occur by assuming an average breakdown field of $500\,\mathrm{kV\,m^{-1}}$ (see Section 3.3), a value of down-conductor inductance of $10^{-6}\,\mathrm{H\,m^{-1}}$, a grounding resistance of 25 ohms (the maximum allowed value recommended by most electrical codes but a difficult value to achieve in sandy soils – see Tables 5.1, 5.2 and 5.3), a maximum current of 100 kA (only a few percent of lightning first strokes will exceed this value), a maximum current derivative of $2.33 \times 10^{11}\,\mathrm{A\,s^{-1}}$ (a high value for both first and subsequent strokes, a value two to three times smaller being typical), and $h=2\,\mathrm{m}$ (about the height of a tall individual). The resistive and inductive components of $V(t)$ will be maximum at slightly different times since the current derivative peak precedes the current peak, as illustrated in Fig. 4.11. The resistive and inductive time-domain voltage waveforms will nevertheless overlap and add. The peak resistive voltage is 2.5×10^6 volts (25 ohms $\times$ 100 kA) and the peak inductive voltage is 0.466×10^6 volts ($10^{-6}\,\mathrm{H\,m^{-1}} \times 2\,\mathrm{m} \times 2.33 \times 10^{11}\,\mathrm{A\,s^{-1}}$). The total peak voltage is 2.62×10^6 volts which occurs before the current has

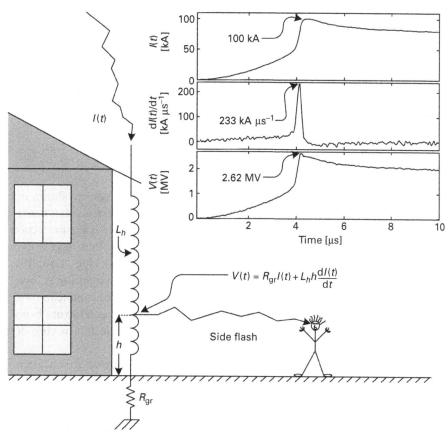

$$V(t) = R_{gr}I(t) + L_h h\frac{dI(t)}{dt}$$

Fig. 4.11 Geometry of a side flash and the currents and voltages involved. Note that the current is plotted as a positive quantity for illustrative purposes (it can represent either a positive or a negative return stroke), whereas for the scientific data in Chapters 1 and 2 the current from negative strokes is plotted as a negative quantity to indicate that negative charge is lowered to ground, as is the convention in much of the literature.

reached peak value. Thus a side flash distance of roughly 5 m length, primarily due to the ground resistance, will occur for the parameters assumed. If the ground resistance were zero, the inductive voltage could produce a side flash of about 0.5 m length. (For very short-duration applied voltages, as is the case for the inductive voltage, the breakdown electric field strength is increased by a factor of about two over the value for a normal impulse voltage breakdown). The potential side flash distance of 5 m shown in Fig. 4.11 will be decreased (1) if the ground resistance is lower than 25 ohms, perhaps naturally made so via ground surface arcing (Figs. 5.5 and 5.6), a close relative of the through-the-air side flash; (2) if there are multiple parallel down conductors, reducing the overall inductance (two down conductors present half the inductance of one down conductor; four, one-quarter the inductance of one); (3) if the height h is less than 2 m; and (4) if the maximum current is the typical value of 30 kA rather than the high value of 100 kA chosen for the calculation (although protection design should involve the extreme cases) and the current derivative is commensurately lower. On the other hand, the ground resistance can be higher than assumed and the breakdown field value depends on the actual geometry of the breakdown gap, so even longer side flashes are possible in principle.

4.5   Very long down conductors

The Atlas V and Space Shuttle lightning protection systems shown in Fig. 4.1 and Fig. 4.2, respectively, are examples of protection systems that contain very long down conductors. As noted earlier, another example is the protection system for some utility power lines, an overhead ground wire that represents both the air terminal and part of the down-conductor system (Fig. 1.10, Fig. 12.2, and Fig. 12.3). The remainder of the down-conductor system for an overhead power line can be either a series of metal towers or a series of conducting wires extending down wood or cement utility poles with all down conductors being electrically connected to grounding electrodes. Power-line protection is considered in more detail in Chapter 12.

In protection systems with very long down conductors, special attention must be paid to the voltage insulation level between the air terminal or down conductors and the object to be protected. When rapidly increasing lightning current is injected into, say, the top of the protective wires in Fig. 4.1 and Fig. 4.2, which are insulated from the poles supporting them and are grounded hundreds of meters away, the voltage at the current injection point rises with the current and is not "grounded" until a reverse (opposite polarity) voltage wave reflects back from the ground to the injection point. No propagating current or voltage wave can travel faster than the speed of light (3×10^8 m s^{-1} or 300 m μs^{-1}). So, for example, if the down conductor from the current injection point to the ground were 150 m, the signal would take a minimum of 1 μs to travel the 300 m round trip. During that microsecond the voltage at the injection point would rise to a value given by multiplying the current at 1 μs – which might be the peak current or a significant fraction of the peak current – by one-half the "surge impedance" of the down conductor, typically of the order of 500 ohms. So, for a current of 30 kA, the voltage at the strike point could reach a maximum of 7.5×10^6 volts before the "relief" wave of opposite voltage arrives. For shorter down conductors, the ground relief signal would arrive sooner, cutting off the rising voltage at a lower level.

It is clear from the above that the voltage standoff level of the insulation, a fiberglass pole or poles in the cases shown in Figs. 4.1 and 4.2, has to be sufficient to withstand a voltage that depends on the length of the down conductors as well as the rise time and peak value of the lightning current. To make the protection design even more uncertain, little is known about the time necessary for a long electrical flashover to develop across a long fiberglass pole or through air to the protected vehicle. Thus very high voltages might be present (and acceptable) for very short times, perhaps a microsecond or so, without a flashover resulting. In Figs. 4.1 and 4.2, the down conductors are about 300 m in length from the insulating fiberglass poles to the ground, and the heights of fiberglass poles used in these configurations and in the previous Saturn V/Apollo protection system are 15 to 22.5 m (50 to 75 feet). The average electric field intensity to cause long spark breakdown in non-uniform meter-length laboratory gaps in air is roughly $500 \, \text{kV m}^{-1}$ for negative voltage applied to the rod and $300 \, \text{kV m}^{-1}$ for a positive rod, if the voltage rise time is near one microsecond (see Section 3.3). Many subsequent strokes will reach peak current within the round-trip down-conductor time of 2 µs; most first strokes will reach perhaps a third to a half of their peak current in 2 µs given their longer current rise times (see Table 2.1 and Fig. 2.1). It follows that the 7.5×10^6 volts calculated in the previous paragraph for an average first stroke current (without canceling voltage wave) can marginally be insulated by a 15 m (50 foot) fiberglass pole. The short duration of such high voltage owing to the arrival of a canceling voltage wave or waves from the ground is likely to be significant in suppressing a complete flashover of the fiberglass pole. According to Fig. 4.3, and other similar video events, the protection system works, although low-level, short-duration electrical discharges along the fiberglass pole would probably not be captured by the video.

For normal lightning protection systems like those on residential structures, the down conductors are generally shorter than about 10 m, and hence the ground relief wave (reflected from the grounding resistance) arrives at the strike point in a very small fraction (less than 1/10) of a microsecond, limiting the rise of voltage at the strike point if the grounding is good. As noted earlier, the voltage at the strike point and on the protection system can still be very large if the grounding resistance is not well controlled. We discuss grounding in the next chapter.

References

Berger, K. and Vogelsanger, E. 1966. Photographische Blitzuntersuchungen der Jahre 1955–1965 auf dem Monte San Salvatore. *Bull. Schweiz Elektrotech. Ver.* **57**: 599–620.

Berger, K. and Vogelsanger, E. 1969. New results of lightning observations. In *Planetary Electrodynamics*, ed. S. C. Coroniti and J. Hughes. New York: Gordon and Breach, pp. 489–510.

Berger, K., Anderson, R. B. and Kroninger, H. 1975. Parameters of lightning flashes. *Electra* **80**: 23–37.

Binczewski, G. J. 1995. The point on a monument: a history of the aluminum cap of the Washington Monument. *JOM* **47** (11): 20–25 (JOM is a publication of The Minerals, Metals and Materials Society).

Dix, E. H. Jr. 1934. Aluminum cap piece on Washington Monument. *Metal Progress* (December), 32–34.

Franklin, B. 1753. Benjamin Franklin, Poor Richard's Almanac for 1753. In *The Papers of Benjamin Franklin*, ed. L. W. Labaree, B. Wilcox, A. Lopez *et al.* Vol. 1, dated 1959, to Vol. 31, dated 1995. New Haven, CT: Yale University Press.

Golde, R. H. 1968. Protection of structures against lightning. *Proc. IEE* **115**: 1523–1529.

IEC 62305-1:2006. *Protection Against Lightning. Part 1: General Principles.* Geneva: International Electrotechnical Commission.

IEC 62305-2:2006. *Protection Against Lightning. Part 2: Risk Management.* Geneva: International Electrotechnical Commission.

IEC 62305-3:2006. *Protection Against Lightning. Part 3: Physical Damage to Structures and Life Hazard.* Geneva: International Electrotechnical Commission.

IEC 62305-4:2006. *Protection Against Lightning. Part 4: Electrical and Electronic Systems Within Structures.* Geneva: International Electrotechnical Commission.

IEC 62305-5:2006. *Protection Against Lightning. Part 5: Services* (to be published). Geneva: International Electrotechnical Commission.

McEachron, K. B. 1939. Lightning to the Empire State Building. *J. Franklin Inst.* **227**: 147–217.

Moore, C. B., Aulich, G. D. and Rison, W. 2000a. Measurement of lightning rod response to nearby strikes. *Geophys. Res. Lett.* **27**: 1487–1490.

Moore, C. B., Rison, W., Mathis, J. and Aulich, G. 2000b. Lightning rod improvement studies. *J. Appl. Meteorol.* **39**: 593–609.

NFPA 780:2004. *Standard for the Installation of Lightning Protection Systems.* Quincy, MA: National Fire Protection Association.

Uman, M. A. and Rakov, V. A. 2002. A critical review of non-conventional approaches to lightning protection. *Bull. Am. Meteorol. Soc.* **83**: 1809–1820.

Yokoyama, S., Miyake, K. and Suzuki, T. 1990 Winter lightning on Japan sea coast: development of measuring systems on progressing feature of lightning discharge. *IEEE Trans. Power Delivery* **5**: 1418–1425.

5 Structure protection: grounding

5.1 Overview

The primary purpose of a lightning grounding system is to provide a means to direct lightning current from the down conductors into the Earth while minimizing the voltage rise on the protection system. For example, if a relatively high peak current I_p of 100 kA is injected into a grounding electrode with a resistance R_{gr} of 25 ohms, the peak voltage V_p on the metal components of the above-ground protection system due to the current flowing through the resistance R_{gr} will be $V_p = R_{gr} \times I_p$, and hence will be equal to 2.5×10^6 volts (see Sections 2.3 and 4.4). This voltage level will lead to side flashes from the above-ground system to any isolated (or grounded at a distance) conducting bodies (metallic or human) within about 5 m of the protection system, since the average electric field for breakdown in air is about $500 \, \text{kV} \, \text{m}^{-1}$ (see Section 3.3). To eliminate such side flashes and the danger to individuals near the protection system, one should bond (electrically connect) nearby metallic objects to the lightning protection system conductors. Since half of the peak lightning currents for first strokes are larger than about 30 kA and 90 percent are larger than about 10 kA, and since it is generally impractical to obtain grounding resistances below about 10 ohms except in the very best conducting soils, the lowest expected peak voltage on a down-conductor system will generally be hundreds of thousands of volts.

Far and away the most common grounding electrode is the vertical ground rod. NFPA 780:2004 specifies a minimum length of 2.4 m (8 feet) for the buried rod. According to that document, the bottom of the ground rod should extend at least to a depth of 3 m (10 feet), which for rods of less than 3 m length requires the top of the ground rod to be buried below ground level. Rod diameters specified in various standards generally equal or exceed 5/8 inch (1.59 cm) for steel rods and 1/2 inch (1.27 cm) for copper rods or copper-clad steel rods, although we shall see in Section 5.3 that the exact rod diameter is not particularly critical to the resultant grounding resistance. Ground rods are generally pounded (driven) into the ground with special hammer-like devices. Long rods with relatively small diameters or sequences of vertically connected rods with relatively small diameters may be more difficult to drive than rods with larger diameters, particularly in dense soil. IEC 62305-3:2006 allows that the minimum length specified for a ground rod, essentially the same as in NFPA 780:2004, may be disregarded if the grounding resistance of the rod is 10 ohms or less (see Section 5.3).

A buried-wire ring electrode (also called a counterpoise) encircling the protected structure is a less common grounding electrode than the ground rod but generally provides a lower grounding resistance. A buried metal mesh beneath the structure (or both a mesh and a counterpoise, bonded together) provides the optimum grounding configuration for a typical structure. Figure 5.1 shows both a trench being dug in preparation for laying a ring electrode and the metal reinforcing bars (rebar) in the structure's foundation, over which concrete is poured, that will form a grounding mesh. A ring electrode should be placed at a distance of about 1 to 2 m from the structure being protected, preferably where water running off the roof can wet the soil around the buried wire (see Section 5.2), and the counterpoise should be buried as deep as is practical. NFPA 780:2004 specifies that the depth of the counterpoise should be greater than 0.46 m (18 inches). NFPA 780:2004 also states that a ring electrode may be encased in concrete if the concrete is in direct contact with the Earth (see Section 5.2). The concrete can be part of the structure foundation. If so, the ring electrode is likely not to be outside the roof edge, as recommended above for a directly buried counterpoise. Different locations on a ring electrode can be used to bond metallic service grounds, although it is preferable for all service grounds to be bonded at one point on the ring to assure that no inductive voltage differences occur between the grounds of different services (see Section 2.3, Fig. 2.4). A ring electrode has the advantage, relative to a ground rod, of tending to equalize the voltages at all points within the ring (assuming no significant inductive voltage differences occur) by way of the principle of electrostatics that states that if the electrostatic potential (voltage) is equal at all points on a closed surface, the potential has the same value within the surface in the absence of electrical charges within the surface. (This principle is exact only for a 3D closed surface, but the 2D ring at constant potential provides an approximation to this situation on the plane within the ring.) Thus, achieving a low grounding resistance is not as critical when a ring electrode is used for grounding since the likelihood of internal side flashes is minimized when voltage differences are minimized. If two or more ground rods are used on a given structure, as is typical, the individual rods should be electrically connected by a bare, buried wire, essentially a piece of a ring electrode, or optimally, a complete ring electrode.

As noted above, perhaps the best grounding system for an ordinary structure is a metal mesh buried (in earth or concrete) beneath the structure. This is so both because of the mesh's relatively low grounding resistance and because of the minimization of voltage differences between different points on the mesh due to the relatively low inductance of the multiple paths between those points. The addition of a bonded counterpoise further improves the grounding and decreases the probability of side flashes, as noted above. Grounding meshes are standard, for example, in utility substations where equalization of voltage differences is essential. When building a house on a concrete slab, typical of most home construction in Florida, the rebar elements in the foundation should be well bonded to provide an effective grounding mesh. Electrical connection points should be made available from that grounding mesh to bond to other grounding electrodes. Examples

Fig. 5.1   A trench being dug to lay a ring electrode (counterpoise). The metal reinforcing bars (rebar) in the building foundation (left) are visible and form a metal mesh that is to be bonded to the ring electrode before cement is poured over the rebar. Courtesy of Bonded Lightning Protection, Inc.

include the ground rod commonly installed by the telephone company to serve as the ground for its required surge arrester (located outside the structure at the point of entry of the telephone wires; see Sections 6.3 and 7.3) and the rod driven by the power utility company as the ground at its meter box (located on the exterior of the structure where the power lines enter). Preferably, all service grounds should be physically located at the same point and be bonded to the grounding mesh at that point, for the reason described in Section 2.3 and illustrated in Fig. 2.4, although the inductive voltage differences associated with a grounding mesh are generally less than those associated with a counterpoise.

5.2   Soil properties

As we shall show in Section 5.3, grounding resistances of buried metal electrodes are directly proportional to the value of the soil "resistivity" where the electrodes are buried. The resistivity is a fundamental physical characteristic of any conductor, including the Earth. According to Saraoja (1977), the local soil resistivity ρ depends largely on the water content of the soil and the resistivity of that water, the relation being given by Hummel's empirical formula:

$$\rho = \left(\frac{1.5}{p} - 0.5 \right) \rho_v \tag{5.1}$$

where ρ is the soil resistivity in ohm-meters, ρ_v is the resistivity of the water in the soil in ohm-meters, and p is the relative volume of water in the soil. If $p = 0.1$ then $\rho = 14.5\rho_v$. From Eq. (5.1), if $p = 0$, ρ would be infinite (an insulator), so the equation fails in this limit although the resistivity of perfectly dry soil is indeed very high. Since the soil resistivity varies according to the content and characteristics of the water in the soil, knowledge of the resistivity of natural waters also gives some insight into the issue of soil resistivities. Table 5.1 gives the resistivities of representative waters and soils. Note that the reciprocal of the soil resistivity is the soil conductivity, $\sigma = 1/\rho$, and hence any of the formulas in this book containing the resistivity can be written in terms of the conductivity.

Completely dry concrete has a very high resistivity, but when concrete is embedded in the ground, moisture penetrates into it and its resistivity becomes about the same as that of the surrounding soil (Saraoja 1977). For this reason, well-bonded rebar or ring electrodes in concrete foundations make relatively good ground electrodes. Well-bonded rebar in buried vertical concrete support pillars for towers and other significant structures also provides a good method of grounding.

The resistivity of the Earth can be reduced by adding chemicals to the soil surrounding the grounding electrode to make the soil more conducting. These commercially available chemicals are either poured onto the soil, combined with soil dug from the hole in which the electrode is to reside and then replaced in the hole, or released by special hollow ground rods that are filled with materials that

Table 5.1 Water and soil resistivities.

Types of water or soil	Resistivity in ohm-meters
Water in oceans	0.1 to 0.5
Sea water at the coasts of Finland	1 to 5
Ground water, well, and spring water	10 to 150
Lake and river water	100 to 400
Rain water	800 to 1,300
Commercial distilled water	1000 to 4000
Chemically clean water	250 000
Clay	25 to 70
Sandy clay	40 to 300
Peat, marsh soil, and cultivated soil	50 to 250
Sand	1000 to 3000
Moraine	1000 to 10 000
Ose (calcereous remains)	3000 to 30 000

Adapted from Saraoja (1977).

diffuse through holes in the surface of the hollow rods. The degree of improvement obtained is not documented in the reviewed literature, although the approaches appear reasonable.

5.3 Resistance of various grounding electrodes

Calculation of the grounding resistance of a given piece of buried metal (ground electrode) in the dc or low-frequency case is a straightforward exercise in electrostatics. We can illustrate the approach to this calculation by considering a geometrically simple grounding electrode, a conducting hemisphere of radius r_0 whose flat surface is flush with the surface of the Earth, as shown in Figure 5.2. The soil resistivity is ρ (in ohm-meters) and is assumed to be homogeneous. If a lightning current I is injected into the electrode, the magnitude of the current density $\mathbf{j}$ in the Earth, that is, the current per unit area perpendicular to the direction of current flow, assuming that current flows uniformly radially outward from the hemispherical surface, is

$$\mathbf{j} = \frac{I}{2\pi r^2} \mathbf{a}_r, r > r_0 \tag{5.2}$$

where r is the radial distance from the strike point at the center of the circular flat surface of the hemisphere, $2\pi r^2$ represents the underground surface area of an imaginary hemisphere at radial distance r through which the current must flow, and the boldface indicates a vector quantity ($\mathbf{a}_r$ is a vector of magnitude unity pointing in the radial direction at any angle from the strike point – see Fig. 5.2). Ohm's Law in "point" form mathematically relates the current density at a point in

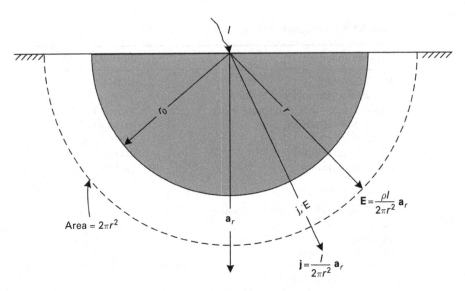

Fig. 5.2   The calculation of the grounding resistance of a metal hemisphere.

a conducting material to the electric field intensity **E** at that point in the material, where **E** and **j** are both in the radial direction

$$E = \rho\,j \tag{5.3}$$

That is, the electric field provides the force that drives the electrons that form the current density, and the proportionality constant between the field and the current density is the resistivity. If we combine Eqs. (5.2) and (5.3), we find that the electric field can be expressed as

$$\mathbf{E} = \frac{\rho I}{2\pi r^2}\mathbf{a}_r, r \geq r_0 \tag{5.4}$$

As illustrated in Fig. 5.3, the difference in the voltage between any two points in the Earth or on the Earth's surface at two different radii, $r = a$ and $r = b$, is found from electrostatic principles as

$$V_{ab} = -\int_b^a E_r\,dr = \frac{\rho I}{2\pi}\left(\frac{1}{a} - \frac{1}{b}\right), \quad a > r_0,\ b > a \tag{5.5}$$

where $\int_b^a$ is a symbol indicating the mathematical function of integration from b to a, and V_{ab} is the voltage at radius a with respect to the voltage at radius b. The voltage of the surface of the metal hemisphere (or any point on the equipotential conductor) with respect to a distant point is found by setting a in Eq. (5.1) to the hemisphere radius r_0 and b to some value much larger, say 100 times larger, or, equivalently, setting b equal to infinity (∞), so that

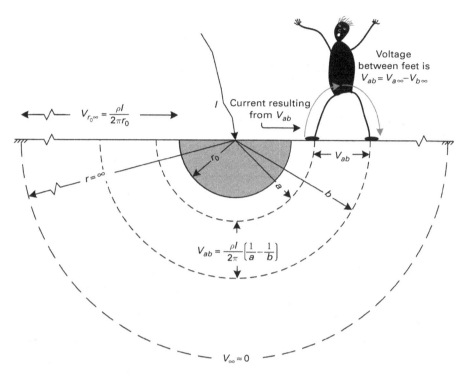

Fig. 5.3 The mechanism of step voltage for current injected into the metal hemisphere of Fig. 5.2. The dashed lines represent surfaces of equal voltage (potential), called equipotential surfaces.

$$V_{r_0\infty} \cong \frac{\rho I}{2\pi r_0} \tag{5.6}$$

The grounding resistance of the metal hemisphere is found by dividing the voltage of the ground electrode given by Eq. (5.6) by the current flowing into the ground electrode that causes the voltage, that is, by applying Ohm's Law in bulk form

$$R_{gr} = \frac{V_{r_0\infty}}{I} \cong \frac{\rho}{2\pi r_0} \tag{5.7}$$

Thus the grounding resistance of the metal hemisphere is directly proportional to the ground resistivity and inversely proportional to the radius of the hemispherical grounding electrode. It follows that the resistance of any grounding electrode is lowered both by increasing the size of the electrode and by lowering the resistivity of the soil where the electrode is buried.

The voltage difference between any two points on the Earth's surface at different radii a and b is called the "step voltage" because it will appear between the two feet of a standing human with one foot at a and the other at b, as shown in Fig. 5.3, when lightning current flows uniformly outward in the soil from the strike point. The step

Table 5.2 Grounding resistance of a typical ground rod of 1/2 inch diameter (1.27 cm) as a function of rod length for two ground resistivities. The ground resistivity of 100 ohm-meters represents a relatively well-conducting soil whereas 1000 ohm-meters represents a moderately poorly conducting soil (see Table 5.1). The rod diameter of 0.5 inch is the minimum allowed by NFPA 780.

Length l	$\rho = 100$ ohm-meters	$\rho = 1000$ ohm-meters
3 m (about 10 feet)	35 ohms	350 ohms
6 m (about 20 feet)	22 ohms	220 ohms
9 m (about 30 feet)	15 ohms	150 ohms

voltage is given by Eq. (5.5) to the extent that the lightning strike-point can be treated as a small hemisphere, not an unreasonable approximation. The higher the soil resistivity and the current, the higher the step voltage between a and b; the larger the distance between a and b, the larger the voltage between a and b. Step voltages cause shocks to humans and animals and can kill a four-legged animal standing near a struck object, for example a tree, because the current flows in the front legs and out the back legs traversing the heart of the animal. There is more discussion about step voltage in Section 7.4 where a typical step voltage is calculated.

As we have noted in Section 5.1, the most common grounding electrode is the vertical ground rod. The resistance of a vertical ground rod, according to Dwight (1936), is

$$R_{gr} = \frac{\rho}{2\pi l}\left(\ln\frac{8l}{d} - 1\right)$$

(5.8)

where l is the length of the rod in contact with the soil, d is the rod diameter, and ln is the natural logarithm (a mathematical function found on many hand calculators). The term $\ln(8l/d) - 1$ is relatively insensitive to variations in l and d compared with the term ρ/l. Thus, the grounding resistance depends primarily on the length of the rod in the ground (the longer the rod, the less the resistance) for a fixed ground resistivity, and on the ground resistivity (a lower resistivity yields a lower resistance) for a fixed rod length. The grounding resistance is relatively insensitive to the diameter of the rod. Table 5.2 lists the ground rod resistance for three rod lengths and for two ground resistivities: moderately good, 100 ohm-meters, and moderately bad, 1000 ohm-meters (see Table 5.1). One important reason to use relatively long ground rods, besides the lowering of grounding resistance apparent from Table 5.2 and Eq. (5.8), is that the soil resistivity often decreases with depth because deeper regions of soil are often wetter, further lowering the grounding resistance of the rod. The combined resistance of two identical rods bonded (electrically connected) together at their tops is roughly half that of either rod individually if the rods are farther apart horizontally than their length. Four such connected rods, for example on the four corners of a structure, presents one-quarter

Table 5.3 Low-frequency ground resistance of a buried horizontal wire of 1/2 inch (1.27 cm) diameter buried at a depth of 0.5 m (about 20 inches) as a function of the length of the wire for two ground resistivities. A ground resistivity of 100 ohm-meters represents a relatively well-conducting soil whereas 1000 ohm-meters represents a moderately poorly conducting soil (see Table 5.1).

Length l	$\rho = 100$ ohm-meters	$\rho = 1000$ ohm-meters
50 m (about 165 feet)	4.0 ohms	40 ohms
100 m (about 330 feet)	2.6 ohms	26 ohms
200 m (about 660 feet)	1.4 ohms	14 ohms

the resistance of an individual rod. In such an arrangement, the connecting wires should be bare and buried, to lower even further the overall grounding resistance of the system.

Expressions for the grounding resistances of a variety of grounding electrodes of different shapes are given by Saraoja (1977). An expression for the low-frequency grounding resistance of a buried horizontal conductor such as a long straight wire or a ring electrode (counterpoise) encircling a structure is given in Eq. (5.9) and, like the vertical ground rod, is primarily dependent on ρ/l where l is the total length of the wire.

$$R_{gr} = \frac{\rho}{\pi l}\left(\ln\frac{2l}{\sqrt{ad}} - 1\right) \tag{5.9}$$

In Eq. (5.9) the wire radius is a, and the wire is buried at depth d. Table 5.3 contains some resistance values derived from Eq. (5.9). Since a buried horizontal conductor of some tens of meters length is often easier to install than a vertical rod of similar length (usually installed as a sequence of connected vertical rods), buried horizontal electrodes are often preferable from a practical point of view. But, as noted above, a vertical ground rod may extend downward into an area of lower soil resistivity than exists near the surface, perhaps below the water table, thus providing a lower grounding resistance than a horizontal conductor of similar length buried near the surface.

In our discussion of grounding above, we have assumed that the grounding electrode impedance is resistive. This is a good approximation for currents containing frequencies under 1 kHz or so (varying on a millisecond scale or slower, such as lightning continuing currents). For transient currents containing frequencies above 100 kHz (such as a return stroke current having microsecond or submicrosecond rise time, Fig. 2.1), it is generally necessary to take account of propagating waves on the grounding system, or otherwise take account of the inductances and capacitances of the grounding system, if the grounding system is more than a few tens of meters long. At its origin a grounding system appears to be a transmission line characterized by a surge impedance (the ratio of voltage to current at the origin) until the time that the first reflection from its far end arrives back at the origin, and such an extended

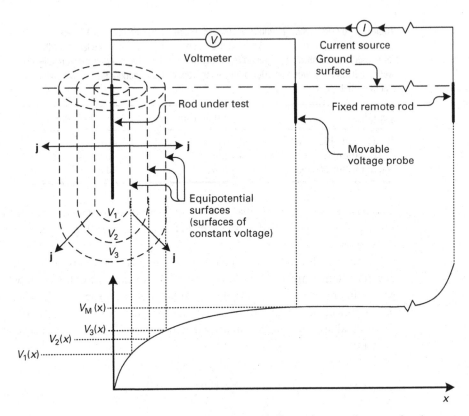

Fig. 5.4   The fall-of-potential method for measuring the grounding resistance of a ground rod.

grounding system after many continuing reflections can be viewed as a simple electrical circuit with inductance, capacitance, and resistance. Further, for high lightning currents the soil will suffer electrical breakdown, that is, the ground connection point will no longer follow Ohm's Law, and hence the grounding resistance will be a non-linear function of the amplitude of the current flowing. Both non-resistive and non-linear grounding electrodes are discussed further in Section 5.4.

Sunde (1968) and many other authors have discussed the techniques for measuring low-frequency grounding resistance and for measuring soil resistivity. The same instrument can be used to measure both. The fall-of-potential method for measuring grounding resistance is illustrated in Fig. 5.4, which also illustrates the geometry of the equipotential surfaces surrounding a ground rod, in contrast to the hemispherical equipotential surfaces surrounding the hemispherical grounding electrode shown in Fig. 5.2 and Fig. 5.3. In Fig. 5.4, current is injected into the rod under test and returns to the generator through the Earth via the fixed remote rod. The injected current in commercially available systems is oscillatory at a frequency higher than 50 Hz or 60 Hz, so as to be immune from power frequency interference. The current generator must produce a voltage large enough to drive the current through the soil and across the interface between each rod and the soil. A voltage amplitude near 50 volts is usually sufficient. The voltage between the rod under test and the movable voltage probe is measured as a function of distance between those two rods. When the voltage between the rod under test and the movable probe has little variation with increasing distance of the movable voltage probe, the rod resistance can be found as V_M/I. The region surrounding the grounding electrode under test should not contain substantial extraneous buried conductors since these conductors will cause a measurement error. For the configuration illustrated in Fig. 5.4 in which the resistance of a single rod is determined, the distance between the rod under test and the fixed remote rod is typically near 35 m, and the voltage vs. distance curve would be optimally flat near 20 m from the electrode under test, that is, about 60 percent of the distance from the rod under test to the fixed remote rod. The 60 percent number is typically used if only a single measurement with the movable probe voltage is desired. If the grounding resistance of an extended grounding system, such as a counterpoise surrounding a structure, is to be determined, the distance between that system and the fixed remote rod must be comparably larger than indicated above. Commercial instruments directly display grounding electrode resistance or soil resistivity when the probes are properly planted.

5.4 Non-resistive and non-linear grounding electrodes

In the calculations of grounding resistance presented in Section 5.3, it is assumed that Ohm's Law in point form applies and that the Earth resistivity is uniform. If the electric field at the grounding electrode surface, given by Eq. (5.4), exceeds the soil breakdown value, typically in the range 100 to 500 kV m^{-1} (Petropoulos 1948, Liew and Darveniza 1974, Mousa 1994), electrical breakdown (arcing) will occur from the electrode both across the Earth's surface and into the surrounding soil, lowering the grounding resistance by effectively enlarging the size of the electrode. For the hemispherical grounding conductor considered in Section 5.3 there will be ionization in the Earth within a radius r_{bd} ("bd" for breakdown) surrounding the grounding conductor where

$$r_{bd} = \sqrt{\frac{\rho I_p}{2\pi E_b}}, \quad r_{bd} > r_0 \tag{5.10}$$

and E_b is the soil breakdown electric field. Equation (5.10) is derived by setting E in Eq. (5.4) to E_b and solving for the maximum radius of the breakdown field. For example, if peak current $I_p = 30$ kA, $\rho = 1000$ ohm-meters, and $E_b = 300$ kV m^{-1}, r_{bd} is about 4 m. It is likely that thermal instabilities will cause the diffuse current in any initially ionized volume to collapse into one or more localized arcs, each carrying a significant portion of the total current. That is, if one region becomes randomly hotter, its resistivity will be lowered, and the region will absorb more of the total current and get even hotter, the pattern continuing until the region contains the entire current flow. Evidence of such soil breakdown in the laboratory is shown in Fig. 5.5 and from lightning damage to a golf course green in Fig. 5.6.

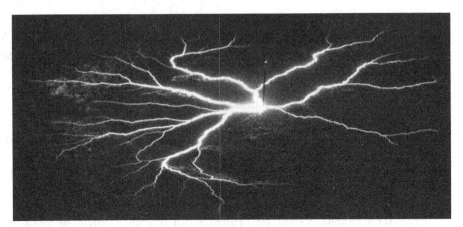

Fig. 5.5   Photograph of surface arcing of about 4 m radius from the point of current injection via a ground rod into soil in a laboratory experiment (Wang *et al.* 2005). Courtesy of Liew Ah Choy.

Fig. 5.6   Photograph showing evidence of electrical breakdown across the Earth's surface from the current of natural lightning injected into a grounding electrode, in this case the pole on a golf course green. Courtesy of E. Philip Krider.

In the laboratory experiment shown in Fig. 5.5, the soil was wet loamy sand of 270 ohm-meters resistivity and was sprayed by artificial rain. The electrical parameters that produced the visible surface arcing and presumably unseen underground arcing were: an input peak current of about 20 kA, a voltage at current peak of about 150 kV, an initial peak voltage (prior to the current peak) of about 230 kV (which initiated the arcing), and a rise time to peak current of 5 to 10 µs. Perhaps the best laboratory studies of electrical discharges in the soil are by Liew and Darveniza (1974), Wang *et al.* (2005), and Sekioka *et al.* (2006). From high-voltage laboratory experiments and from model calculations, they show that electrical breakdown of the soil around a ground rod significantly reduces the dc grounding resistance in a few microseconds, a time comparable to the rise time of the applied current in the experiments. Soils of higher resistivities have greater decreases in the ratio of the breakdown-reduced resistance to the initial resistance. As an example, for an injected current above 50 kA and an initial ground rod resistance of 300 ohms in a soil of resistivity 1000 ohm-meters, the 300 ohm resistance is reduced by about a factor of 5, to 60 ohms, via the arcing associated with the high current flow, whereas an initial 30 ohm ground rod resistance in a soil of resistivity 100 ohm-meters is reduced by only about a factor of 2, to 15 ohms. Nor (2006) presents a bibliography of 48 publications concerned with the electrical characteristics of soil subjected to large impulse currents.

We now examine the issue of the transient or high-frequency response of grounding electrodes in situations where the grounding system is relatively large. When transient current is injected into an extended earth electrode, the electrode initially exhibits a surge (or characteristic) impedance (not a grounding resistance), which for a long buried horizontal wire is 150 to 200 ohms (Bewley 1963). The transient signal propagates along the buried conductor, generally with a speed about one-third the speed of light (less than the speed of light in air because of the relatively high permittivity of the Earth and the ohmic losses associated with the soil resistivity), and reflects back and forth between the ends of the electrode while being damped by radial current flow into the Earth (Bewley 1963). The initial voltage increase with time at the source results from the summation of (1) the rising voltage at the source due to the rising current at the source multiplied by the surge impedance and (2) the reflected, opposite-polarity voltage wave arriving from the end of the electrode, in about 1 µs for an electrode of 50 m length. If the dc (low-frequency) value of the grounding resistance of a long buried wire is less than its surge impedance, as is generally the case, the input impedance at the source is reduced to its dc value after a few reflections. For a 50 m counterpoise the initial source impedance will be the surge impedance, reducing in a few microseconds to a resistance value near the dc or low-frequency value of Eq. (5.9), numerical examples of which are shown in Table 5.3. A high initial voltage may lead to electrical breakdown within or across the surface of the soil at the source (see previous paragraph), assuming the rising current causes that high value before the "relief" voltage reflection arrives. Since the lightning current (particularly the current of a subsequent stroke) can rise to its peak value in under 1 µs, buried horizontal wires much longer than 50 m may not provide a significant advantage over shorter wires, at least for the first microseconds of the lightning current. NFPA 780:2004 specifies that buried horizontal ground electrodes or ground ring electrodes, whether buried in concrete or not, should always exceed 6 m (20 feet) in length.

References

Bewley, L. V. 1963. *Traveling Waves on Transmission Systems*. New York: Dover Publications, Inc.

Dwight, H. B. 1936. Calculation of resistances to ground. *Trans. Am. Inst. Electr. Eng.* **55**: 1319–1328.

IEC 62305-3:2006. *Protection Against Lightning*. Part 3: *Physical Damage to Structures and Life Hazard*. Geneva: International Electrotechnical Commission.

Liew, A. C. and Darveniza, M. 1974. Dynamic model of impulse characteristics of concentrated earths. *Proc. IEE* **121**: 123–125.

Mousa, A. M. 1994. The soil ionization gradient associated with discharge of high currents into concentrated electrodes. *IEEE Trans. Power Delivery* **9**: 1669–1677.

NFPA 780. *Standard for the Installation of Lightning Protection Systems*, 2004 edn. Quincy, MA: National Fire Protection Association.

Nor, N. M. 2006. Review: soil electrical characteristics under high impulse currents. *IEEE Trans. Electromagn. Compatibility* **48**: 826–829.

Petropoulos, G. M. 1948. The high voltage characteristics of earth resistances. *JIEEE* **95**: 59–70.

Saraoja, E. K. 1977. Lightning earths. In *Lightning*, Vol. 2, *Lightning Protection*, ed. R. H. Golde. New York: Academic, pp. 577–598.

Sekioka, S., Lorentzou, M. I., Philippakou, M. P. and Prousalidis, J. M. 2006. Current dependent grounding resistance model based on energy balance of soil ionization. *IEEE Trans. Power Delivery* **21**: 194–201.

Sunde, E. D. 1968. *Earth Conduction Effects in Transmission Systems*. New York: Dover Publications, Inc.

Wang, J., Liew, A. C. and Darveniza, M. 2005. Extension of dynamic model of impulse behavior of concentrated grounds at high current. *IEEE Trans. Power Delivery* **20**: 2160–2165.

6 Surge protection for electronics in low-voltage electrical systems

6.1 Overview

As noted in Section 1.4 and Section 3.2, it is common to consider separately (1) the lightning protection of a structure and (2) the lightning protection of the electrical power, electronic equipment (e.g., television, DVD, burglar alarm system, computer), and communication systems (e.g., telephone, cable television) located within that structure, although, as we have discussed in Section 3.1, it is preferable to consider the two aspects of protection in a unified way as part of an overall topologically shielded and surge-protected system. The electrical power, electronic equipment, and communication systems within a structure are generally connected to the outside world by conducting wires, primarily 50 Hz or 60 Hz utility power wiring and telephone cables. Outside metallic wires are exposed to direct lightning strikes and suffer significant induced voltages and currents due to lightning occurring with a few hundred meters of those wires. In this chapter we consider primarily the protection of low-voltage systems. By "low voltage" we mean any voltage equal to or below the normal household power level, 480 volts generally being the highest "low voltage" satisfying this definition. In Section 12.2 we will examine the use of surge arresters on distribution and transmission power lines where the voltages may vary from some thousands of volts on distribution lines to about a million volts on transmission lines.

Surge protective devices (SPDs) are known by a number of different names including "arresters," "surge arresters," and "lightning arresters." Their purpose is to limit the lightning-induced transient overvoltage and divert the associated surge current, usually to protect electrical or electronic equipment from being damaged. On utility power lines and some other applications, SPDs are also used to inhibit flashover (electrical breakdown) between a conductor raised to a relatively high voltage by lightning and adjacent conductors at lower voltage. Additionally, SPDs provide protection from transient voltages and currents caused by sources other than lightning, examples including the transients that occur when electric motors in air conditioners turn on or off and the disturbances in the external utility system called "switching surges" that occur when power company loads are connected or disconnected. Motors and transformers, which have many closely spaced turns of wire, and equipment containing low-voltage solid state electronics are particularly susceptible to damage from transient overvoltages.

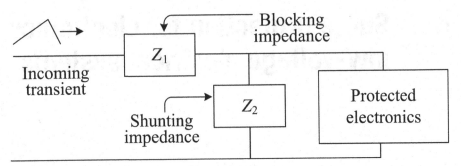

Fig. 6.1   A generic lightning protection system for electronic equipment. Adapted from Standler (2002).

A generic transient protection system for electronic equipment is shown in Fig. 6.1. In Fig. 6.1, the circuit impedance Z_1 is intended to provide a "blocking" function and the impedance Z_2 is intended to provide a "shunting" function for the unwanted transients. The impedance Z_1 can be a resistor, an inductor (if the frequency content of the transient is higher than that of the signal), or a capacitor (if the frequency content of the transient is lower than that of the signal). The impedance Z_2 is generally an SPD, but may be a circuit element that forms a low-pass or high-pass filter with Z_1. Under normal operating conditions (no transients), Z_1 should approximate a short circuit or low impedance in series with the circuit being protected (to appear as if it is not present), and Z_2 should approximate an open circuit or high impedance in parallel with the circuit being protected (again to have no influence on normal circuit operation). If this is not the case during normal operation, the transient protection could adversely distort the normal operating signal. As an example, SPDs in parallel (Z_2) can act as unintended capacitors, providing a low impedance at higher frequencies (the impedance of a capacitor is inversely proportional to the operating frequency) and reducing the amplitude of and otherwise distorting the normal operating signal, a situation that can be particularly troublesome.

An excellent general reference for the material contained in this chapter is the book by Standler (2002). Additionally, the reader is referred to the book by Hasse (2000), the standard from the International Electrotechnical Commission IEC 62305-4:2006, and IEEE standards IEEE C62.41.1:2002, IEEE C62.41.2:2002, and IEEE C62.45:2002.

6.2   Amplitudes and waveshapes of lightning transients

In order to design protective circuits to reduce the potentially harmful effects of lightning transients, one needs to know the expected waveforms and maximum signal amplitudes of the transients. From measurements made on communication lines and on utility power lines, both outside and within structures, "standard"

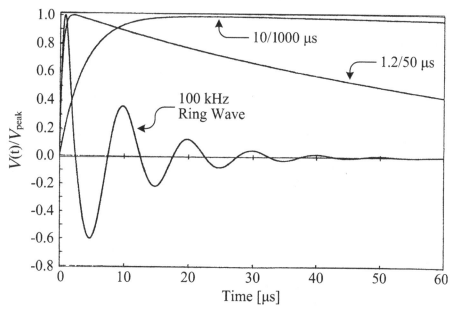

Fig. 6.2   Three different overvoltage test waveforms: (1) the 10/1000 μs waveform for testing the lightning protection (immunity) of communications circuits; (2) the 1.2/50 μs voltage waveform and (3) the 100 kHz ring waveform for testing the lightning protection (immunity) of low-voltage power systems. Adapted from Standler (2002).

waveforms have been derived for designing protective systems. Nevertheless, any given lightning strike, particularly if it is very close, may produce a "non-standard" transient waveform.

Communication lines generally have a different exposure to lightning than power lines. Communication lines are often mounted beneath distribution power lines, and communication lines are usually shielded and bundled. Based on the measurements of Bodle and Gresh (1961), reasonable worst-case overvoltages to which telephone lines are exposed can be expected have a rise time to peak value of 10 μs and a time to decrease to half of peak value of 1000 μs. Such a transient is called a 10/1000 μs waveform. This voltage waveform does not represent a typical transient, but rather a composite worst case specified for testing the lightning immunity of communication systems. Figure 6.2 shows a plot of the first 60 μs of the 10/1000 μs waveform. The 10/1000 μs waveform is found in a number of standards, as are similar waveforms such as the 10/700 μs. One standard for gas tube arresters (see Section 6.3) calls for the 10/1000 μs waveform with a peak current of 500 amperes. The peak voltage/peak current found in other representative standards is 1500 volts/40 amperes, 800 volts/100 amperes, and 1500 volts/200 amperes.

While studying the failure of household electric clock motors, Martzloff and Hahn (1970) found that for a clock motor insulation level of 6000 volts, the motor failure rate was 1 percent of the failure rate that occurred with a 2000 volt insulation level. The implication of this observation is that there are few transients on the 60 Hz 120 volt service with a maximum value greater than 6000 volts, but there are a considerable number of transients greater than 2000 volts. Because of the spacing between wires and other metal components, the insulation level in household electrical wall outlets and circuit breaker boxes is generally between 5000 and 10 000 volts. Thus one would not expect to observe higher level transients inside a structure because of prior electrical breakdown at these weak points (which act as spark-gap crowbar arresters: Section 6.3). In other studies of transients on power lines inside structures, it was found that most transients recorded on 120 volt lines were in the range of several hundreds of volts with the upper 1 percent level in different studies being between 300 volts and 2000 volts. Lightning transients in the thousand-volt range on a structure's 120 volt lines are probably due to lightning within a few hundred meters of the structure, an event that occurs several times a year to most structures in the United States (Sections 1.2 and 1.4).

The most common waveforms specified in standards for designing lightning protection of low-voltage power lines within structures (as well as for designing protection for high-voltage equipment on distribution and transmission lines) are the 8/20 μs current waveform and the 1.2/50 μs voltage waveform. The 1.2/50 μs voltage waveform is shown in Fig. 6.2. The 8/20 μs current waveform and the 1.2/50 μs voltage waveforms are intended to approximate the direct effects of a lightning first return stroke which is typically responsible for the largest magnitude transient during a lightning discharge. Note, however, that the rise time of the current in subsequent return strokes (return strokes following the first stroke) is generally less than 1 μs (Section 1.3; Fig. 1.9, Fig. 2.1, Fig. 2.2), and thus this rapidly changing transient signal is not accounted for in these standard waveforms. Within structures, the maximum voltage specified for testing is generally about 6000 volts (because, as noted above, other elements of the power system will suffer electrical breakdown first, preventing higher voltages), and the maximum current is about 3000 amperes. In addition to the unipolar voltage waveforms shown in Fig. 6.2, a so-called ring waveform, based on the measurement of Martzloff and Hahn (1970), is specified in some standards. The ring waveform of Martzloff and Hahn (there are other ring waveforms) is also shown in Fig. 6.2. Martzloff and Hahn observed that transient overvoltages on low-voltage systems often have a rise time to peak that is less than 1 μs, followed by a decaying oscillation at a frequency of about 100 kHz. One standard sets the peak current for the ring waveform at 500 amperes and the peak voltage at 6000 volts.

Individual SPDs are rated by the maximum continuous operating voltage (rated voltage) they can withstand without failure, the total transient power and energy they can withstand without failure, the peak current they can withstand without failure, and the peak voltage that will appear across the arrester terminals when a current of a certain magnitude and waveform is passed through the arrester. We discuss some of the individual types of SPDs in the next three sections, Sections 6.3–6.5. In Section 6.6 we consider the use of multiple SPDs in protection circuits.

6.3   Crowbar devices

When transient voltage of sufficient amplitude is impressed across an SPD crowbar device, that device will suffer an internal electrical breakdown which reduces the voltage across the terminals of the device to typically a few volts. Crowbar devices include air spark gaps (two conducting electrodes separated by atmospheric air), gas tubes (two metal electrodes separated by a low-pressure gas), or solid-state crowbar devices such as thyristors, silicon-controlled rectifiers, and triacs. The *V–I* (voltage vs. current) characteristic of an idealized gas tube crowbar device is given in Fig. 6.3. The device is inherently bipolar; that is, it does not matter which polarity of voltage is applied. In the example shown, very little current flows through the gas tube (it remains an insulator) until the voltage across the tube increases to a value above about 600 volts. At that voltage, electrical breakdown occurs in the tube, and the resulting discharge transitions through a "glow" region to the "arc" region. In the arc region, the designed operating region, a relatively small voltage exists across the tube for a wide range of currents, between about 1 and 10 000 amperes. Gas tubes are usually not appropriate on dc (direct current) power circuits because the dc current, once initiated, is difficult to turn off, and hence the short-circuit current flow could damage the dc power source and/or destroy the SPD. Gas tubes are used in some ac (alternating current) power circuits where the design is such that the resulting 50 Hz or 60 Hz arc extinguishes following a zero value in the current waveform (as the current oscillates from one polarity to the other), restoring the device to its previous open-circuit state.

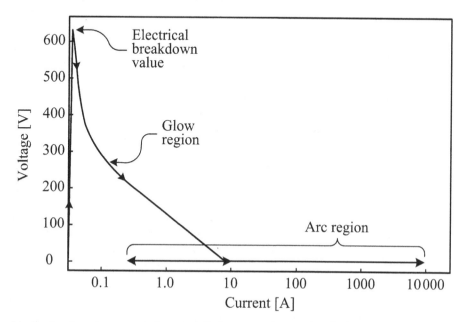

Fig. 6.3   Idealized voltage vs. current characteristic for a gas tube crowbar device.

The oldest, simplest, and least expensive crowbar device is the carbon-block arrester. This device is still used today in some telephone line protection, but it is rapidly being replaced by sealed, gas-filled tubes with metal electrodes inside. Carbon-block arresters have atmospheric air between two carbon electrodes. The electrodes are separated by a fraction of a millimeter so they will suffer electrical breakdown when a voltage over about 600 volts is applied. Repeated firing or the passage of high currents causes the overall gap to widen, raising the breakdown voltage, and often leaves carbon particles free to move in the gap, resulting in telephone noise. These negative effects are reduced by the use of sealed spark gaps. In general, spark gaps have two primary advantageous features: (1) they can conduct large current, tens of thousands of amperes for tens of microseconds, without degrading their mode of operation, and (2) they have the smallest capacitance, typically about 1 picofarad, of any SPD and thus are one of the few protective devices that will not interfere with proper circuit behavior in the radio frequency range above about 50 MHz. Disadvantages of spark gaps are that they can be relatively slow to operate (break down electrically and transition through the glow discharge to the arc mode), difficult to turn off if current continues to flow after the transient has ended (as noted above), require a relatively large voltage to turn on, and operate at a relatively low voltage when conducting (which can be either an advantage or a disadvantage). If a voltage rising to 1000 volts in 1 μs is applied to a typical spark gap arrester with dc breakdown voltage of about 500 volts, the turn-on (firing) time will be about 1 μs. If 500 volts is applied in 1 μs to the same arrester, the turn-on time will be about twice as long; and for 5000 volts applied in 1 μs to the same arrester, the firing time will be a few tenths of a microsecond. Clearly, a significant voltage, at the dc breakdown voltage or above, will bypass the spark gap arrester during the turn-on time of tenths of a microsecond to several microseconds.

Silicon-controlled rectifiers (SCRs), triacs, and other types of thyristors are solid-state crowbar devices that have similar advantages and disadvantages to spark gaps. They can handle large currents but are relatively slow to turn on and possibly difficult to turn off. The SCR and triac are three-terminal devices in which a "gate" terminal switches the current on and off through the other two terminals. SCRs only conduct in one direction, while triacs conduct in either direction.

6.4   Voltage clamping devices

Voltage clamping devices such as metal oxide varistors (MOVs), Zener (avalanche) diodes, and p–n junction diodes attempt to hold the device or circuit input voltage at a near-constant value somewhat above the normal operating voltage of the protected device or circuit while conducting and diverting transient currents. In general, semiconductor junction diodes are used in electronic circuits with operating voltage levels up to about 10 volts, whereas MOVs are available that can provide clamping voltages from a few volts to the million volt levels of power transmission lines. MOVs are manufactured in sizes ranging from small disks the size of a fingernail for use on

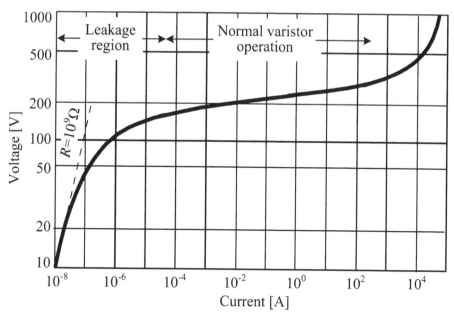

Fig. 6.4 Voltage vs. current characteristic for a metal-oxide varistor (MOV) used on a 110 volt household circuit. Adapted from Littelfuse arrester catalogue AN 9767.1.

printed circuit boards to disks nearly 10 cm (about 4 inches) in diameter that can be stacked (electrically connected in series) to build an overall MOV of several meters length for use on power transmission lines. Protective circuits containing both clamping and crowbar devices, as opposed to the use of either separately, combine the positive features of both devices and will be discussed in Section 6.6.

Voltage clamps have the advantage that they can operate in nanoseconds (10^{-9} s) or less. They generally have the disadvantage of relatively large capacitance (in the nanofarad range) making them problematic for radio frequency circuits. MOVs can be constructed to conduct relatively large currents, up to tens of thousands of amperes. Diodes can only carry relatively small currents, tenths of an ampere to 100 amperes or so. Junction diodes clamp between 0.7 and 2 volts, avalanche diodes between about 6.8 and 200 volts.

A typical V–I (voltage vs. current) curve for an MOV is shown in Figure 6.4. In the example shown, intended for a household 120 volt, 60 Hz service, the voltage is maintained at a value near 200 volts for a current range from less than 10^{-3} amperes to near 10 kA. MOVs are basically non-linear resistors. Ohm's Law, $V = R \times I$ (Section 2.3), is not strictly valid since it requires R to be a constant value. Nevertheless, it can be used if R is allowed to change as a function of the current flowing through it (or the voltage across it). Thus, an effective value of R can be found for any point on the curve in Fig. 6.4 by dividing the value of V at that point by the value of I. At low system-operating currents, that resistance is very high, as illustrated in Fig. 6.4, so that an MOV used as Z_2 in Fig. 6.1 will not interfere with

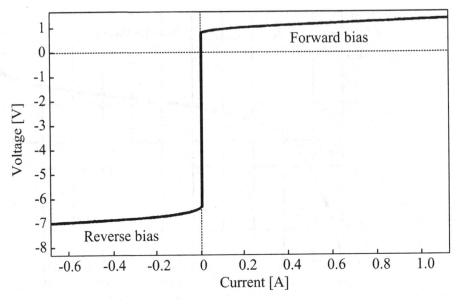

Fig. 6.5   Typical voltage vs. current characteristic for an avalanche diode.

normal circuit operation. "Varistor" stands for "voltage-variable resistor." The "metal oxide" in the acronym MOV is mostly zinc oxide. MOVs are bipolar. They behave the same way (same V–I curve) no matter which polarity of voltage is applied or which direction the current flows in the device.

A typical V–I curve for an avalanche diode is shown in Fig. 6.5. In the forward-biased direction (right-hand side of Fig. 6.5), the diode behaves like an ordinary rectifier diode, providing a clamping voltage in the 1 volt range for a wide range of current. In the reverse-biased direction (left-hand side of the figure), where the voltage across the diode is negative, the diode clamps at a voltage near 7 volts, via avalanche behavior. The device with characteristic shown in Fig. 6.5 would generally be used as an avalanche diode. Normal forward-biased semiconductor diodes are used for clamps when protection in the 1 volt range is needed. Note that all diodes intended for use in the reverse breakdown region are commonly called Zener diodes, although, strictly speaking, the Zener mechanism only operates to about 5 volts. Avalanche diodes, operating via the avalanche mechanism, are available with breakdown voltages up to the hundred volt range.

6.5   Filters and isolation devices

Electrical circuit filters, with or without non-linear circuit components, can be used to absorb and reflect unwanted transients while passing the desired signals. Filters generally operate by blocking some range of frequencies in the incoming signal waveform, and passing others. For example, in ac power circuit protection, all

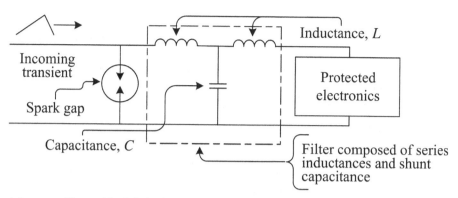

Fig. 6.6 A low-pass filter to block lightning transients and pass power frequencies (50 Hz and 60 Hz) preceded by a spark gap crowbar device.

frequencies above a few hundred hertz could be blocked (a low-pass filter) and those frequencies below a few hundred hertz passed, effectively suppressing lightning transients with frequency content generally in the 10^3 Hz to 10^6 Hz range, but leaving the 50 or 60 Hz operating signal unaffected. Simple filters can be constructed with series inductors (Z_1 in Fig. 6.1) and shunt capacitors (Z_2 in Fig. 6.1). Such filters may be damaged by large transient signals so they are often used in combination with a crowbar or clamping SPD, as illustrated in Fig. 6.6. One interesting form of series inductance, providing a relatively large blocking impedance to the higher-frequency unwanted transients, consists of "ferrite beads," high permeability magnetic material that can be clamped on flat cable or threaded on individual wires.

In most of the discussion thus far in this chapter, SPDs have been assumed to be acting on an undesirable transient applied between a signal wire and a ground wire. In some circumstance, both the signal wire and the ground wire can be simultaneously raised to high voltage by lightning. This so-called "common mode" transient, as contrasted with the "differential mode" transients we have been primarily considering, can be protected against by isolation devices called isolation transformers and optical isolators (optoisolators). These devices have no direct conducting path between input and output, and thus they block any transient that presents the same undesirable voltages on the signal and on the ground wire. Isolation transformers use a magnetic field to couple between input and output; optoisolators use a light signal. Both devices are designed to block unwanted common-mode transients to values of about 5000 volts, but neither blocks differential-mode transients, so additional differential-mode protection is needed.

Isolation transformers are similar to ordinary transformers, generally used to raise or lower power frequency (50 Hz or 60 Hz) voltages, but isolation transformers have grounded electrostatic shielding between the primary and secondary coils in an attempt to eliminate parasitic capacitive coupling between the coils that can transfer the common-mode voltage to the secondary coil.

Optical isolators use a light-emitting-diode or a laser to convert the incoming electrical signal to a light signal that propagates inside transparent plastic fibers to a photo-detector. Small optoisolators are available for printed circuit boards and longer fiber optic runs are used between separate pieces of equipment. The fiber optic cable is also immune to induced voltages and currents caused by the electric and magnetic fields of lightning, whereas multiple metallic-wire cable and coaxial cable can have unwanted voltages and currents induced by those fields.

6.6   Hybrid circuits: multiple-stage protection

The use of a spark gap arrester in combination with a filter to form a hybrid protection circuit is illustrated in Fig. 6.6. As another example of a hybrid protection circuit, the use of a spark gap arrester in combination with a voltage clamping device such as an MOV is illustrated in Fig. 6.7. With proper design, the decoupling inductor and resistor shown in Fig. 6.7 between the two SPDs allow the spark gap to operate first and absorb or reflect the bulk of the input energy so the MOV does not reach its voltage, current, energy, or power limit, and fail. The decoupling impedance is usually an inductor for power systems and usually a resistor for data or telecommunication systems. The initial part of the transient voltage is primarily impressed across the spark gap and the decoupling impedance so that little energy reaches the MOV until the spark gap fires. Clearly, knowledge of the transient waveform (see Section 6.2) is critical in determining the value of the decoupling inductance required, since the sum of the voltages across the inductor, $L \, dI/dt$, and the resistor, $R \times I$, is the difference between the voltage across the crowbar device and the voltage across the MOV. At relatively high initial rates of change of current, the spark gap will always operate first, absorbing or reflecting the bulk of the surge energy. At relatively low rates of change of current (perhaps not the lightning transient case being anticipated) the MOV may reach its voltage limit first because of the relatively low voltage drop across the inductor. In this case, the spark gap may not fire, and perhaps the MOV will absorb more energy than intended and will fail.

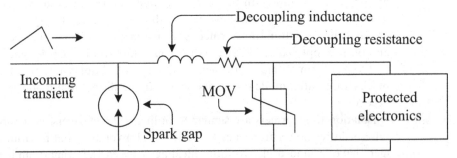

Fig. 6.7   A hybrid protection circuit composed of a spark gap followed by a metal-oxide varistor, with decoupling impedance located between these two SPDs.

Multiple stages of SPDs (more than the two shown in Fig. 6.6 and Fig. 6.7) may be used to obtain additional protection for devices or circuits. The first stage in such a protection system is generally a crowbar device or an MOV that can handle the largest signal expected. The various succeeding stages must be decoupled by series impedances. The voltage clamping level at the final stage is that necessary to protect the sensitive device or circuit. Such networks can be designed theoretically, but it is always best to test them in the laboratory to make sure they can withstand the range of transient signal amplitudes and durations expected (see Section 6.2). As noted earlier, these voltage and current waveforms are different for communication networks than for power networks. As also previously noted, lightning does not always produce waveforms similar to those specified in the standards.

If a number of SPDs are used for power system protection in separated locations within one structure, such as one at the entrance of the primary power to a residence and additional SPDs at the power input to televisions and computers within the residence, the voltage clamping levels of all the SPDs should be about the same, or preferably, a lower clamping voltage should be used for the primary, heavy duty MOV at the structure entrance and a somewhat higher clamping voltage should be used for the secondary MOVs on household electronic devices (but not high enough to allow the devices to be damaged). On a 120 V, 60 Hz circuit, the primary SPD should typically be rated at 180 V (similar to the MOV V–I characteristic in Fig. 6.4), with the secondary SPDs at that level or higher. If this coordination of SPD clamping voltage levels is not properly considered, the SPD with the lowest rated voltage (perhaps at or within an electronic device such as a hi-fi amplifier) may operate first and be destroyed as it attempts to "protect" the higher-voltage-related SPDs in the system. This will occur if the decoupling impedances provided by the inherent inductance (typically about 10^{-6} H m^{-1}) of the power cables throughout the structure do not adequately decouple the SPDs. Note that these decoupling inductances, which allow the primary SPD to operate first on the incoming transient, are uncontrolled (not designed), but, nevertheless, they are often fortuitously adequate for the intended purpose. The primary SPD should be located on the outside of the structure, at the service entrance. Primary SPDs are commercially available to fit in an empty circuit-breaker slot in the circuit-breaker panel through which the primary structure power passes. NFPA 780:2004 states that this primary SPD should have a maximum current rating greater than 40 kA, applied as a 8/20 μs waveform, per incoming phase, and that on 120 volt lines the maximum value of suppressed voltage should be 500 volts, while on 240 volt lines it should be 1 kV.

References

Bodle, D. W. and Gresh, P. A. 1961. Lightning surges in paired telephone cable facilities. *J. Bell Syst. Tech.* **40**: 547–576.

Hasse, P. 2000. *Overvoltage Protection of Low Voltage Systems*, 2nd Edn. London: The Institution of Electrical Engineers.

IEEE C62.41.1:2002. *IEEE Guide on the Surge Environment in Low Voltage (1000 V or less) AC Power Circuits*. New York: IEEE.

IEEE C62.41.2:2002. *IEEE Recommended Practice on Characterization of Surges in Low Voltage (1000 V or less) AC Power Circuits*. New York: IEEE.

IEEE C62.45:2002. *IEEE Recommended Practice on Surge Testing for Equipment Connected to Low-Voltage (1000 V or less) AC Power Circuits*. New York: IEEE.

IEC 62305-4:2006. *Protection Against Lightning. Part 4: Electrical and Electronic Systems Within Structures, General Principles*. Geneva: International Electrotechnical Commission.

Martzloff, F. D. and Hahn, G. J. 1970. Surge voltages in residential and industrial power circuits. *IEEE Trans. Power Apparatus Syst*. **89**: 1049–1056.

NFPA 780:2004 *Standard for the Installation of Lightning Protection Systems*. Quincy, MA: National Fire Protection Association.

Standler, R. B. 2002. *Protection of Electronic Circuits from Overvoltages*. New York: Dover Publications, Inc.

7 Humans and animals

7.1 Personal safety

An eminent group of lightning safety experts, meeting at the American Meteorological Society's annual conference in 1998, formulated the following guidelines regarding safety from lightning injury and death (Holle *et al.* 1999):

No place is absolutely safe from the lightning threat; however, some places are safer than others.

- Large enclosed structures (substantially constructed buildings) tend to be much safer than smaller or open structures. The risk for lightning injury depends on whether the structure incorporates lightning protection, construction materials used, and the size of the structure.
- In general, fully enclosed metal vehicles such as cars, trucks, buses, vans, fully enclosed farm vehicles, etc., with the windows rolled up provide good shelter from lightning. Avoid contact with metal or conducting surfaces outside or inside the vehicle.
- Avoid being in or near high places and open fields, isolated trees, unprotected gazebos, rain or picnic shelters, baseball dugouts, communications towers, flagpoles, light poles, bleachers (metal or wood), metal fences, convertibles, golf carts, and water (ocean, lakes, swimming pools, rivers, etc.). When inside a building *avoid* use of the telephone, taking a shower, washing your hands, doing dishes, or any contact with conductive surfaces with exposure to the outside such as metal door or window frames, electrical wiring, telephone wiring, cable TV wiring, plumbing, etc.

Where groups of people are involved, an action plan for getting to a lightning-safe place must be made in advance by the responsible individuals.

Zimmerman *et al.* (2002) expanded the above guidelines and added medical information, without otherwise changing the overall recommendations. Walsh *et al.* (2000) published a similar set of lightning safety recommendations particularly focused toward college athletes. Both of these sources recommend and discuss the so-called 30–30 rule for defining the periods of time that are unsafe relative to the lightning threat. The 30–30 rule is considered in Section 8.1.

Many, perhaps even most, individuals struck by lightning live to tell the story, but some suffer long-lasting injuries. Immediate medical treatment, primarily cardiopulmonary resuscitation (CPR), can potentially save a lightning strike victim from dying. It is a common misconception that individuals struck by lightning carry an electrical charge, rendering them hazardous to touch. This is definitely not the case. There is no personal electrical hazard involved in administering CPR to a lightning strike victim.

A bibliography on lightning safety, numbering about 250 publications, can be accessed via request from ron.holle@vaisala.com.

7.2   Statistics

According to the monthly magazine Storm Data, a publication of the US National Oceanographic and Atmospheric Administration (NOAA), 85 lightning-related deaths occurred per year, on average, in the United States in the years from 1966 through 1995. The Storm Data statistics are derived primarily from newspaper clippings describing weather-related injury and death. Distributions of lightning deaths and injuries according to the place of occurrence are displayed in Fig. 7.1. Over 30 percent of all lightning deaths involve individuals who work outdoors and over 25 percent involve outdoor recreationists. Comparison of the Storm Data statistics with information from the state of Florida, state of Texas, and state of Colorado public health and medical records indicates an under-reporting of deaths by Storm Data of roughly 30 percent (Holle *et al.* 2005). In view of this under-reporting of lightning deaths in the national statistics, it is likely that about 100 individuals, on average, are actually killed by lightning in the United States each year.

According to Storm Data, about 300 individuals are injured by lightning in the United States annually. Colorado medical records indicate that there is an under-reporting of about 40 percent for the lightning injuries in Colorado compiled by Storm Data (Lopez *et al.* 1993). If we extrapolate this result to the whole United States, it is likely that more than 500 people are injured, on average, each year. In

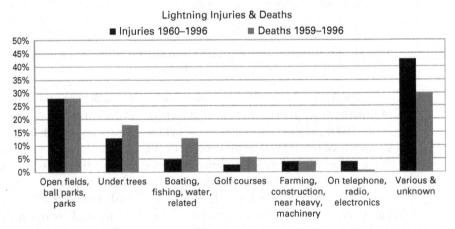

Fig. 7.1   The place of occurrence of lightning deaths and injuries in the United States, 1959–1996. Adapted from Storm Data, US National Climatic Data Center, NOAA, Ashville, North Carolina.

fact, the number could be quite a bit higher if minimal injuries, not usually reported, are included. In the Colorado medical records, the ratio of lightning injury to death is about 10 (Cherington *et al.* 1999), a result implying that about 1000 individuals are injured annually in the United States if 100 are killed.

Storm Data contains a wealth of detailed information on lightning-related fatalities, injuries, and damage. From that data base, Curran *et al.* (2000) have developed many tables and figures containing death, injury, and damage statistics, examples of which are found in Table 7.1 and Table 7.2. For the period 1959 through 1994, Storm Data contains reports of 3239 deaths, 9818 injuries, and 19 814 cases of property damage. Florida led the nation in both the number of deaths, 345, and the number of injuries, 1178. The largest number of damage reports, 1441, came from Pennsylvania. Nationwide, there were 0.42 lightning deaths per million people per year. New Mexico and Wyoming had the two highest death and injury rates per million residents of the state with 1.88 deaths and 13.84 injuries in New Mexico and 1.47 deaths and 5.74 injuries in Wyoming. There was a maximum occurrence of lightning casualties in July, midsummer in the northern hemisphere (in Australia, casualty rates peak in December–January, summer in the southern hemisphere). Two-thirds of the lightning casualties occurred between noon and 4 p.m. local standard time (1 p.m. and 5 p.m. local daylight saving time). Casualties showed a steady increase toward a maximum at 4 p.m. local standard time, followed by a somewhat faster decrease. For incidents involving deaths only, 91 percent of the cases had one fatality, while 68 percent of the injury cases had one injury. Examples of group injuries and deaths are discussed in Section 7.4. Males were five times more likely to be killed or injured than females. There was roughly one lightning fatality for every 90 000 cloud-to-ground flashes in the United States (there are 20 to 30 million cloud-to-ground flashes per year).

Graphs of the annual number of recorded deaths from lightning in the United States and of the US population as a function of time for a period of nearly a century are found in Fig. 7.2. Before about 1920, not all states consistently compiled lightning death records, so there are more deaths than shown in these years. There has been a steady decline in the reported annual death rate since about 1930, even though the population has steadily increased, as is clearly evident from Fig. 7.2. Apparently the decrease in death rate with time parallels a decrease in the percentage of people living in rural areas, that is, the fraction of the population regularly working outdoors in farming and ranching. In addition, the introduction of metallic plumbing and wiring into structures has helped to provide unintended lightning protection for those structures. Finally, as communication and emergency services became more sophisticated, medical aid was able to arrive faster, and there have been advances in resuscitation techniques as well as greater public awareness of and training in those techniques, particularly cardiopulmonary resuscitation (CPR).

The specific locations of lightning-caused deaths in the United States during the period from 1891 to 1894 were compared with similar data from 1991 to 1994 by Holle *et al.* (2005). About one-quarter of the nineteenth century deaths were

Table 7.1 Number of lightning fatalities, injuries, casualties, and damage reports, and their ranks, for states, the District of Columbia, and Puerto Rico from 1959 to 1994.

State	Fatalities No.	Fatalities Rank	Injuries No.	Injuries Rank	Casualties No.	Casualties Rank	Damage reports No.	Damage reports Rank
Alabama	84	16	211	17	295	18	287	28
Alaska	0	51	0	52	0	52	3	52
Arizona	59	24	105	29	164	30	84	43
Arkansas	110	9	245	13	355	12	576	14
California	21	35	58	40	79	38	60	45
Colorado	95	11	299	11	394	10	312	26
Connecticut	13	42	75	35	88	36	269	29
Delaware	15	41	27	44	42	43	83	44
District of Columbia	5	48	18	48	23	49	14	48
Florida	345	1	1178	1	1523	1	450	19
Georgia	81	18	329	10	410	9	656	10
Hawaii	0	52	4	51	4	51	14	49
Idaho	20	37	67	38	87	37	305	27
Illinois	85	15	275	12	360	11	412	21
Indiana	74	22	164	24	238	23	350	24
Iowa	65	23	162	25	227	26	579	13
Kansas	56	25	178	22	234	25	1182	2
Kentucky	82	17	196	19	278	19	566	15
Louisiana	116	6	231	15	347	14	315	25
Maine	22	34	104	30	126	31	253	30
Maryland	116	7	134	26	250	20	455	18
Massachusetts	24	33	331	9	355	13	603	12
Michigan	89	12	643	2	732	2	814	6
Minnesota	53	27	116	28	169	29	406	23
Mississippi	89	13	207	18	296	17	205	33
Missouri	79	20	97	31	176	28	253	31
Montana	20	38	44	42	64	41	88	42
Nebraska	41	30	70	36	111	33	618	11
Nevada	6	47	12	49	18	50	11	50
New Hampshire	8	45	68	37	76	40	206	32
New Jersey	55	26	130	27	185	27	98	41
New Mexico	81	19	168	23	249	21	54	47
New York	128	4	449	5	577	5	1005	3
North Carolina	165	2	464	4	629	4	960	4
North Dakota	11	44	24	45	35	46	145	37
Ohio	115	8	430	6	545	6	412	22
Oklahoma	88	14	243	14	331	15	826	5
Oregon	7	46	19	46	26	48	150	35
Pennsylvania	109	10	535	3	644	3	1441	1
Puerto Rico	30	32	6	50	36	45	4	51
Rhode Island	4	49	45	41	49	42	122	38
South Carolina	77	21	229	16	306	16	717	8
South Dakota	20	39	59	39	79	39	437	20
Tennessee	124	5	349	7	473	8	764	7
Texas	164	3	334	8	498	7	689	9

Table 7.1 (cont.)

State	Fatalities		Injuries		Casualties		Damage reports	
	No.	Rank	No.	Rank	No.	Rank	No.	Rank
Utah	34	31	82	34	116	32	107	39
Vermont	12	43	18	47	30	47	151	34
Virginia	51	28	184	21	235	24	487	17
Washington	3	50	37	43	40	44	56	46
West Virginia	20	40	88	32	108	34	146	35
Wisconsin	47	29	194	20	241	22	509	16
Wyoming	21	36	83	33	104	35	105	40
United States	3239		9818		13057		19814	

Adapted from Curran *et al.* (2000).

indoors, the greatest number of indoor deaths occurring while the victims were sleeping in bed, compared with only a few percent of indoor deaths in the 1990s. In both time periods studied, about 10 percent of deaths occurred when people took shelter under trees, and males accounted for 70 to 80 percent of the deaths.

Lightning fatality statistics are available for a number of countries besides the United States. For example, the average annual number of deaths per million people in the United Kingdom (England, Wales, Scotland, and Northern Ireland) between 1993 and 1999 was 0.05 (Elsom 2001), a factor of about 8 less than in the United States. This is a reasonable ratio given there is much less lightning in the United Kingdom. In all countries for which data are available, there has been a decrease in the annual number of lightning deaths during the course of the twentieth century, no doubt for the same reasons given above for the United States. For example, in the Netherlands more than 20 deaths per year occurred in the 1920s, but since about 1970 there have been only one to five deaths per year (Ten Duis 1998). In England and Wales, the average annual deaths were 19 between 1852 and 1899, 13 for the period 1900 to 1949, and 5 between 1950 and 1999 (Elson 2001). Eighty to ninety percent of the fatalities in England and Wales during these three periods were male, consistent with the statistics from the United States (Curran *et al.* 2000), from Singapore (Pakiam *et al.* 1981), and from Australia (Coates *et al.* 1993). In Singapore there were about 2.6 lightning deaths per year per million population in 1930 and about 1.6 in 1970. In the United States and in Australia the number of deaths per year per million population has decreased from about 1.3 in the 1930s to about 0.3 in the 1980s.

The annual death toll for lightning worldwide is probably a few thousand individuals, as can be crudely inferred from the known worldwide distribution of lightning and the known fatalities in a number of representative countries. The annual worldwide number of serious lightning injuries is probably five to ten times the annual number of deaths, assuming the ratio is similar to that for the United States.

Table 7.2 Average population, and rate per million people per year of lightning fatalities, injuries, casualties, and damage reports, and their ranks, for all states, the District of Columbia, and Puerto Rico from 1959 to 1994. Population is average of decennial census values from 1960 to 1990.

State	Average population [1000s]	Fatality rate Rate	Fatality rate Rank	Injury rate Rate	Injury rate Rank	Casualty rate Rate	Casualty rate Rank	Damage Rate Rate	Damage Rate Rank
Alabama	3660	0.64	24	1.60	23	2.23	22	2.18	34
Alaska	369	0	52	0	52	0	52	0.23	50
Arizona	2364	0.69	19	1.23	30	1.93	26	0.99	44
Arkansas	2086	1.46	3	3.26	4	4.73	4	7.67	7
California	22 275	0.03	49	0.07	51	0.10	50	0.07	51
Colorado	2536	1.04	6	3.24	5	4.28	5	3.42	21
Connecticut	2990	0.12	45	0.70	42	0.82	44	2.50	29
Delaware	564	0.74	17	1.33	27	2.07	24	4.09	17
District of Columbia	691	0.20	44	0.89	37	0.92	41	0.56	45
Florida	8605	1.10	4	3.80	3	4.91	3	1.45	40
Georgia	5119	0.44	28	1.79	17	2.23	23	3.56	19
Hawaii	869	0	51	0.10	49	0.10	51	0.45	46
Idaho	833	0.67	22	2.23	13	2.90	13	10.17	4
Illinois	11 011	0.21	42	0.69	43	0.91	43	1.04	43
Indiana	5223	0.39	29	0.87	38	1.26	37	1.86	35
Iowa	2818	0.64	23	1.60	24	2.24	21	5.71	13
Kansas	2317	0.67	20	2.13	15	2.80	14	14.17	2
Kentucky	3401	0.67	21	1.63	21	2.30	20	4.62	16
Louisiana	3830	0.83	9	1.65	19	2.48	17	2.28	32
Maine	1078	0.57	25	2.68	6	3.25	8	6.52	11
Maryland	4005	0.80	12	0.94	36	1.74	29	3.16	23
Massachusetts	5648	0.12	47	1.63	22	1.75	28	2.97	25
Michigan	8813	0.28	37	2.03	16	2.31	19	2.57	28
Minnesota	3918	0.38	30	0.84	39	1.21	38	2.88	26
Mississippi	2372	1.04	5	2.42	7	3.47	6	2.40	30
Missouri	4758	0.46	27	0.54	44	1.00	40	1.48	38
Montana	739	0.75	15	1.65	20	2.41	18	3.31	22
Nebraska	1511	0.75	16	1.27	28	2.02	25	11.36	3
Nevada	694	0.24	40	0.52	45	0.76	45	0.44	47
New Hampshire	844	0.26	38	2.24	12	2.50	16	6.78	10
New Jersey	7082	0.22	41	0.49	46	0.71	46	0.38	49
New Mexico	1200	1.88	1	3.89	2	5.76	2	1.25	41
New York	17 642	0.20	43	0.71	41	0.91	42	1.58	37
North Carolina	5535	0.84	8	2.32	10	3.16	10	4.82	15
North Dakota	635	0.48	26	1.05	34	1.53	31	6.34	12
Ohio	10 501	0.30	34	1.13	32	1.44	35	1.09	42
Oklahoma	2764	0.88	7	2.42	8	3.31	7	8.30	6
Oregon	2334	0.08	48	0.23	48	0.31	48	1.79	36
Pennsylvania	11 715	0.26	39	1.26	29	1.52	32	3.42	20
Puerto Rico	2296	0.36	31	0.07	50	0.44	47	0.05	52
Rhode Island	939	0.12	46	1.33	25	1.45	34	3.61	18
South Carolina	2895	0.78	13	2.22	14	2.99	12	6.88	9

Table 7.2 (cont.)

State	Average population [1000s]	Fatality rate		Injury rate		Casualty rate		Damage Rate	
		Rate	Rank	Rate	Rank	Rate	Rank	Rate	Rank
South Dakota	683	0.81	10	2.40	9	3.21	9	17.77	1
Tennessee	4240	0.81	11	2.29	11	3.09	11	5.01	14
Texas	12 998	0.35	32	0.71	40	1.06	39	1.47	39
Utah	1283	0.74	18	1.78	18	2.51	I5	2.32	31
Vermont	477	0.76	14	1.05	35	1.80	27	8.79	5
Virginia	5037	0.28	36	1.06	33	1.35	36	2.69	27
Washington	3815	0.02	50	0.27	47	0.29	49	0.41	48
West Virginia	1837	0.30	33	1.33	26	1.63	30	2.21	33
Wisconsin	4492	0.28	35	1.21	31	1.49	33	3.15	24
Wyoming	397	1.47	2	5.74	1	7.21	1	7.35	8
United States	216 738	0.42		1.26		1.67		2.54	

Adapted from Curran *et al.* (2000).

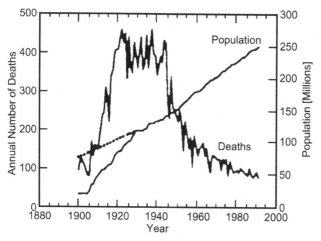

Fig. 7.2   Annual number of lightning deaths, total population of the reporting states (solid line), and total population of the contiguous United States (dotted line and right branch of solid line). Adapted from Lopez and Holle (1998).

7.3   Medical issues

A bibliography of publications on the medical aspects of lightning injury and death is presently found at www.uic.edu/labs/lightninginjury, and an excellent general reference to this Section is the book by Andrews *et al.* (1992). There are a number of ways a human can be killed or injured by lightning: (1) from the current of a direct strike (e.g., Fig. 2.1); (2) from the current of an unconnected upward leader when lightning strikes nearby (e.g., Fig. 1.7b shows three upward leaders, two of which will remain unconnected; Fig. 11.2 shows unconnected upward leaders from the tree and the tower); (3) from a side flash or ground surface arc from a nearby object that is directly struck (Section 4.4, Fig. 4.11, Fig. 5.5, Fig. 5.6); (4) from the step voltage produced by lightning current flowing in the ground from a nearby strike (e.g., Fig. 5.3); (5) from the touch voltage encountered when in contact with a metallic object such as a wire fence that has been raised in voltage by either direct or nearby lightning; (6) from blunt trauma caused, for example, by being thrown many meters by muscle contractions initiated by the lightning strike, a not uncommon occurrence, or, as another example, by being hit by bricks knocked off a building by lightning, or, as a final example, by falling off a horse spooked by lightning; (7) from burns or smoke inhalation from lightning-caused fires; and (8) from lightning-ignited explosions such as occur in underground coal mining operations.

Cooper and Andrews (1995) and Andrews (2006) have categorized lightning effects on humans as (1) minor, (2) moderate, and (3) severe, as discussed below.

(1) In the minor category are those individuals who, when they gain consciousness, report a feeling of having been hit on the head or having been in an explosion. They may or may not remember seeing the lightning or hearing the thunder. They are often confused and amnesic, with temporary deafness, or blindness. They seldom demonstrate any burns or paralysis, but may complain of muscular pain and suffer confusion lasting from hours to days. Their vital signs are usually stable, although victims occasionally demonstrate transient mild high blood pressure. The victims are likely to make a complete recovery.

(2) In the moderate category, individuals may be disoriented, combative, or comatose. They frequently exhibit motor paralysis, with mottling of the skin and diminished or absent pulses, particularly of the lower extremities. Occasionally, these individuals have suffered temporary cardiopulmonary standstill, although it is seldom documented. Spontaneous recovery of a pulse is attributed to the heart's inherent automaticity. Seizures may also occur. First- and second-degree burns may not be immediately prominent but may evolve over the first several hours. Third-degree burns occur rarely. Victims in this category often have at least one eardrum ruptured as a result of the lightning-produced pressure wave (acoustic shock wave) that is the origin of thunder. The victims are likely to recover, but they may exhibit long-lasting sleep disorders, irritability, difficulty with fine psychomotor functions, chronic pain, and general weakness.

(3) In the severe category, victims suffer cardiac arrest with either ventricular standstill or fibrillation. Cardiac resuscitation may not be successful unless it is undertaken immediately. Respiratory arrest may occur and be prolonged leading to secondary cardiac arrest from lack of oxygen. Direct brain damage may occur from the lightning current or possibly from the high pressure of its acoustic shock wave, as well as from a lack of oxygen for too long a period of time because of respiratory or cardiac arrest. Eardrum rupture with bleeding in the middle ear cavity is common. The prognosis is usually poor in this group, but this may be as much due to a delay in administering cardiopulmonary

resuscitation as to the severity of the direct lightning-caused damage. Most fatalities fall in this, the severe, category.

How exactly does lightning kill? It has been argued in the medical literature that the only immediate cause of death in a lightning victim is cardiopulmonary arrest and that a strike victim is highly unlikely to die unless a cardiopulmonary arrest is suffered as an immediate effect of the strike (e.g., Cooper 1980). It has been estimated that, until recently, nearly 75 percent of those who suffered cardiopulmonary arrest from lightning injuries did die, many because cardiopulmonary resuscitation was not attempted. While cardiopulmonary arrest is certainly the primary cause of lightning death, the second cause of death (and injury) is damage to the victim's central nervous system. When current traverses the brain, there may be coagulation of the brain substance, formation of epidural and subdural hematomas, intraventricular hemorrhage, and paralysis of the respiratory center. The most life-threatening of these effects is current flowing in the respiratory control center, leading to respiratory arrest (C. Andrews, M. D. private communication).

It is a commonly held myth that a direct lightning strike victim will be seriously burned, both internally and externally. This is definitely not true. The relatively short duration of the overall lightning current, typically a fraction of a second or less, saves all but a few victims from serious burns. Examples of common types of lightning burns are shown in Figs. 7.3–7.6. Overall, the burns observed in lightning accidents are generally divided into five categories: full-thickness (Fig. 7.3), punctate

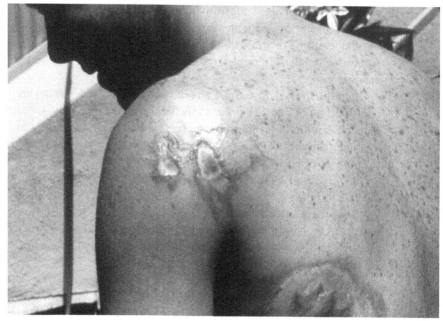

Fig. 7.3   Full thickness skin burns at entrance/exit site. Courtesy of C. Andrews, M. D.

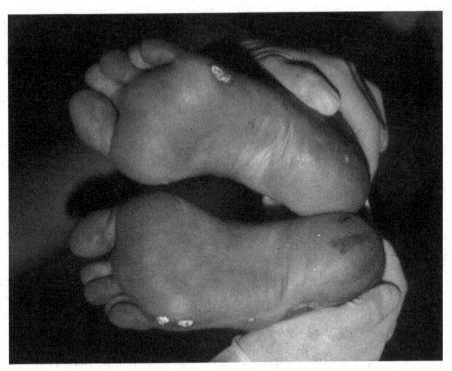

Fig. 7.4   Punctate skin burns on entrance/exit site on feet. Courtesy of C. Andrews, M. D.

(Fig. 7.4), linear (Fig. 7.5), feathering or flowers (also called fern-like patterns, arborescent burns, or Lichtenberg figures) (Fig. 7.6), and thermal burns due to burning clothing or heated and melted metal jewelry, zippers, and belt buckles. Additionally, the ultraviolet radiation from the close lightning channel may produce a type of "sunburn."

Often the shoes and socks of an individual struck by lightning are ripped open and/or blown off the feet, presumably because of the high pressure associated with the vaporization of the moisture confined within the individual's shoes by the hot lightning current. Clothing is also commonly torn and ripped. A strike victim's clothing is displayed on a mannequin in Fig. 7.7.

Many types of lightning-caused eye damage have been reported (e.g., Cannel 1979). Eye injury may be due to (1) direct thermal or electrical effects, (2) intense light including ultraviolet, and (3) the high-pressure acoustic shock wave. Eye injury has been estimated to occur in as many as half of those struck by lightning (Castren and Kytila 1963). Some manifestations of eye injury, for example the development of cataracts, can be long delayed in time.

Temporary deafness is not uncommon in lightning strike victims, likely associated with the acoustic shock wave of close thunder. At least half of lightning strike victims suffer the rupture of one or both eardrums (Cooper 1980). Direct nerve damage from lightning may cause facial palsies.

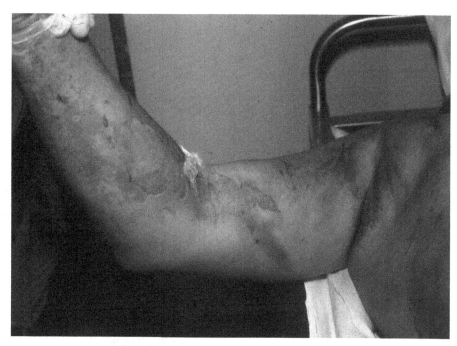

Fig. 7.5   Linear skin burns. Courtesy of C. Andrews, M. D.

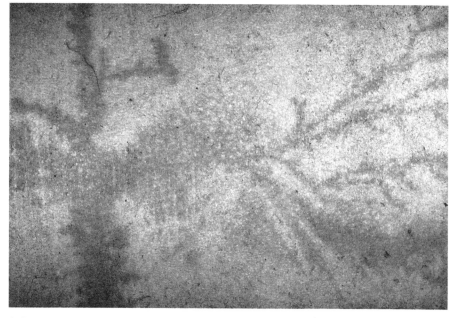

Fig. 7.6   Arborescent skin burns. Courtesy of B. Hocking, M. D.

Fig. 7.7   The torn clothing of a lightning strike victim displayed on a mannequin. Courtesy of Professor M. A. Cooper, M. D. who received the photograph from the victim.

Blumenthal (2005, 2006) has developed guidelines for the autopsies of those killed by lightning, in an attempt to encourage better uniformity in the *post mortem* examination and reporting processes. He states that, in addition to performing the usual complete autopsy, in the case of suspected lightning death the following list should receive special attention:

1. The external examination should include a meticulous description of the clothing and any evidence of attempted resuscitation.
2. Metal objects may have burned the underlying skin, or may have been marked by the heat of electrical arcing. Metal objects may show signs of fusing, zincification, cuprification and/or magnetization. Metal objects such as tooth fillings, spectacles, belts, buckles, coins, and pacemakers should be specifically commented on.
3. The type, pattern, and distribution of any cutaneous thermal injuries should be noted, including clusters of punctate burns, blisters, or charred burns.
4. Rupture of tympanic membranes (use an otoscope) should be noted.
5. Mention should be made of singed and/or scorched hair.
6. Eye signs, such as retinal detachment, should be noted (cataracts can be difficult to demonstrate *post mortem*).
7. Unique arborescent or fern-like injuries (Lichtenberg figures) should be noted.
8. The procedure for internal examination should be identical to that of any careful forensic autopsy.

In addition to the obvious physical effects of lightning on the human body, the psychological effects can be long-lasting and debilitating. In fact, the psychological effects are generally described as more devastating to the victim and capable of producing more of a long-term disability than the obvious physical effects (Cooper 1995a,b; Cooper and Marshburn 2005; see also Cooper's Medical Aspects of Lightning at www.lightningsafety.noaa.gov). Long-term psychological effects, most of which probably have a physical basis in difficult-to-identify nervous system injury, include anxiety, fatigue, chronic headaches or other chronic pain, decreased libido and impotence, personality changes, and depression.

Only a few percent of lightning deaths and injuries occur indoors (see Fig. 7.1). These deaths and injuries are generally from transient voltages associated with cloud-to-ground lightning that terminates on or near an outdoor conductor that enters a structure. Examples of such conductors are telephone wires, metallic plumbing, and power system wires. The transient voltage transmitted inside the structure can result in shock or in side-flashing to an individual touching or near the indoor conductor that has been raised to high voltage. There is considerable literature on telephone-related injury and death, the most common indoor lightning event. In Australia, about 60 people report telephone injuries annually in a population of 16 million (Andrews and Darveniza 1989). One to three telephone-related lightning deaths and an unknown number of injuries occur annually in the United States. In the Australian study, about 35 percent of those injured on the telephone reported hearing a loud noise and feeling an electric shock. Twenty percent reported sparks emanating from the telephone, and 40 percent reported being physically thrown. In the United Kingdom from 1993 to 1999, there was an average of about 25 lightning incidents inside buildings per year, with one-quarter of the indoor incidents involving the telephone (Elsom 2001).

In most countries, telephone signal wires are required by law to be protected from overvoltages caused either by lightning or by contact of the telephone line with a power line. This overvoltage protection is generally accomplished by gas-tube spark gap arresters or carbon-block air gap arresters (see Section 6.3, Fig. 6.3) that reduce the incoming voltage to a small value after it initially exceeds 500 to 1000 V. These SPDs are generally placed between the signal wires and a grounding system (typically a ground rod) at the point that the wires enter a structure. Unfortunately, significant voltages may still appear on telephone handset wiring if the grounding resistance of the SPD is not sufficiently low, the voltage depending on the product of the grounding resistance and the current produced by lightning that flows into the grounding resistance ($V = R \times I$, Section 2.3). Voltages above the 5000 to 10 000 volt range will exceed the insulation level of most telephones, allowing sparks to exit the phone. For example, a relatively small current of 100 amperes flowing through the telephone-protecting crowbar arrester into a ground rod with a grounding resistance of 50 ohms will raise the voltage on both the telephone signal wire and ground wire (these voltages are about the same because the voltage across the crowbar arrester is relatively small) to near 5000 volts (50 ohms × 100 amperes). Voltage differences between the telephone ground and the power and other grounds, including metallic plumbing, should be minimized by bonding, ideally at a common entrance point for all utilities. Otherwise, a situation such as depicted in Fig. 2.4 may occur. Using a cordless or a cellular phone allows one to avoid the potential electrical hazard associated with a hard-wired phone. In addition to electrical injury caused by voltage differences between the various telephone wires in the handset and the human body, relatively large voltage differences between the wires inside the telephone can produce an acoustic shock from the acoustic transducer of the telephone. Acoustic shocks can also occur in cordless telephones. Telephone handsets contain SPDs that are intended to limit the voltages inside the handset and hence limit any acoustic shock, but this approach is not always successful.

Humans are not often killed by step voltages (see Fig. 5.3), step voltages being discussed further in Section 7.4, apparently because the lightning current does not easily find its way to life-critical areas of the human body. Interestingly, four-legged animals, particularly those who happened to be seeking shelter from rain under trees struck by lightning, can have fatal current forced through their hearts and through critical parts of their nervous systems by step voltages that appear between their front and hind legs. There are many examples of groups of animals (e.g., cows, horses, sheep, elk) being killed simultaneously, often without visible lightning effects on their bodies, typically when they have taken shelter under a tree or have been near a fence. A photograph of one such group kill is found in Fig. 7.8. Animals, like humans, can also be killed by direct strikes, unconnected upward leaders, side flashes, surface arcs, touch voltage, blunt trauma, and

Fig. 7.8   A photograph of 23 Holstein heifers killed by lightning near a fence in Island Park, Idaho. Courtesy of Ruth and Kent Bateman.

lightning-caused fire and smoke, the latter being not uncommon for horses confined to stables struck by lightning.

7.4   Electrical effects

The consensus view of the sequence of electrical events occurring when an individual is struck by lightning was first advanced by Berger (1971a,b) and has been reviewed and expanded upon by Golde (1973) and by Andrews *et al.* (1992). As we will see, the full lightning current flows through the body only briefly and at relatively low levels before a surface flashover across the body occurs, shunting most of the lightning current to the outside of the body. The surface flashover reduces the damage that would have occurred to the body if all current had flowed internally.

In order to understand the interaction of lightning currents and voltages with the human body, let us assume that lightning is in the process of striking an individual and that current has just begun to flow through the individual's body, between the head and feet. The current is increasing with time, as shown in the current waveforms of Fig. 7.9 (see also Fig. 1.8, Fig. 1.9, Fig. 2.1). Figure 7.9a illustrates the current through the body and the voltage across the body during the upward-connecting leader phase (see Section 1.3, Fig. 1.6, Fig. 1.7b). That phase is assumed to have a duration of about 1 ms during which time the body current increases from zero to 100 amperes and the voltage across the body from zero to 70 kV. Figure 7.9b shows the electrical situation during the first stage of the return stroke when the current through the individual's body rises from a value of 100 amperes at the end of the upward-connecting leader phase to a magnitude of 1000 amperes in a time of

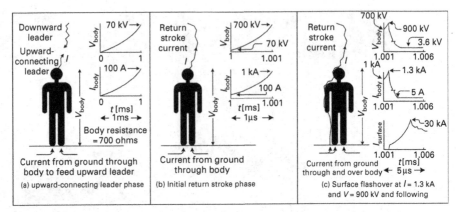

Fig. 7.9   Electrical aspects of a lightning strike to a human: (a) upward-connecting leader phase, (b) initial return stroke phase, and (c) surface flashover and thereafter.

about 1 μs. If we assume that an air gap the height of a human experiences electrical breakdown at an average electric field intensity of about $500\,kV\,m^{-1}$ (see Section 3.3), a flashover through the air across the surface of a human of height, say, 1.8 m (about 6 feet) will be expected when the voltage between the individual's head and ground reaches about 900 kV ($500\,kV\,m^{-1} \times 1.8\,m$). If the surface of the struck person is wet, and this is always the case to some extent, the assumed average electrical field intensity for breakdown may be lower by up to a factor of about two, similarly lowering the value of breakdown voltage from 900 kV to a smaller value. If the individual's body resistance is assumed to be 700 ohms (actual values may be up to a factor of two larger or smaller according to Andrews *et al.* 1992), a flashover at 900 kV will occur when the lightning current through the body is about 1300 amperes (900 kV/700 ohms); that is, at that value of current through the body for which the voltage between head and feet reaches 900 kV, as shown in Fig. 7.9c. If the breakdown voltage is smaller, the associated body current for breakdown will also be smaller for the same assumed body resistance. In Fig. 7.9c the body current increases from the 1000 amperes occurring at the end of Fig. 7.9b to 1300 amperes, at which current level the breakdown occurs across the body surface, lowering the voltage across the body and the current through the body, as discussed in the next paragraph. Note that the surface flashover across the body occurs relatively early in the return stroke current's rise to peak value (the typical peak value is tens of thousands of amperes).

Once surface flashover has taken place, the electric field along the arc that has formed through the air over the surface of the body is about $2\,kV\,m^{-1}$, characteristic of any long arc in air and more or less independent of the arc current, as determined from laboratory arc experiments (e.g., King 1962). Thus, the surface flashover reduces the voltage from head to foot from 900 kV to a much lower value of 3.6 kV ($2\,kV\,m^{-1} \times 1.8\,m$) in a time of a microsecond or so, and the current flowing inside the individual's body is therefore also reduced from about 1300 amperes to about 5 amperes (3.6 kV/700 ohms), a value which will be maintained as long as there is appreciable lightning current flowing along the outside surface of the body, as illustrated in Fig. 7.9c. It follows from the above that tens of thousands of amperes of the lightning current may flow outside the human body by virtue of the surface flashover while only a few amperes will flow inside the body after the surface flashover has occurred. Thus the surface flashover probably saves many individuals from death. For each stroke in a flash, the interior of the human body will be subjected to a short-duration current pulse with peak value near 1000 amperes followed by a relatively constant current of a few amperes maintained for milliseconds or even longer, up to about 100 milliseconds in the case of the flow of continuing current following a return stroke. The scenario presented is consistent with the typical observations of burn marks on the body surface of lightning strike victims and the melting of their metal jewelry and belt buckles. The current calculated to flow in the body is certainly sufficient, if directed into the heart or to particular parts of the nervous system, to cause either cardiac or respiratory arrest or both. Apparently, the relatively short duration of the lightning current makes such an outcome less likely than with the typically longer-duration accidental exposure to 50 or 60 Hz alternating current of equivalent amplitude. In this regard, there is a limited portion of the cardiac cycle when the heart is most susceptible to electrical damage. With a brief shock, or series of shocks, such as produced by lightning, the chances of transgressing this "vulnerable window" are lessened.

Before the return stroke current occurs, the individual being struck by lightning will be subjected to the current of the upward-propagating connecting leader, part of the attachment process, as illustrated in Fig. 7.9a. It seems reasonable that many more individuals would experience unconnected upward discharges that occur in response to nearby lightning flashes than will be struck directly. Such individuals may well report being struck directly. The currents associated with unconnected upward discharges can be of the order of a hundred amperes and can have a duration up to 1 ms or so (Fig. 7.9a). They can also be smaller and of shorter duration. Anderson *et al.* (2002) present frames from two videos of a soccer match in South Africa showing nine soccer players clasping their heads, falling down, and otherwise showing signs of distress coincident with the occurrence of lightning striking ground outside the stadium in which they were playing. At least four players had only one foot touching the ground at the time of the lightning so the step potential mechanism (see next paragraph) would appear to be ruled out. Thus, unconnected upward leaders from the players' heads are implicated as the likely cause of their injuries. After the lightning, six players reported having headaches and four reported having pain in their legs.

In addition to the effects of direct strikes and unconnected upward discharges, individuals can experience a significant fraction of the total lightning current by way of side flashes through the air and current flow within the ground or along the ground surface (Fig. 4.11, Fig. 5.3, Fig. 5.5, and Fig. 5.6). Lightning current can flow in or along the ground and cause death or injury in two different ways. (1) Uniform current flow in the ground, producing a step voltage between the legs in which life-threatening currents generally do not reach the heart or critical parts of the nervous system of a human, a two-legged animal, or a four-legged animal standing or walking upright on only two legs (Fig. 5.3). Nevertheless, an appreciable electric shock may be delivered. Although the lightning current density in the ground decreases inversely with the square of the distance from the source (Section 5.3, Fig. 5.2, Fig. 5.3), the lightning current flowing in the Earth in the vicinity of a strike to ground can be high enough, particularly for relatively high-resistivity soils (those with resistivity of the order of 1000 ohm-meters or more – Table 5.1), to produce widespread electric shocks to nearby groups of people. (2) Surface arc discharges, photographs of two examples being shown in Fig. 5.5 and Fig. 5.6, depicting the results of surface arcing in the laboratory and on a golf course green, respectively. Surface arcs are similar to side flashes, but the arc current flows along the surface of the Earth rather than through the air. Thus injury or death from these surface arcs might well be viewed as due to a form of side flash. Encounters with surface arcs can lead to thermal burns, shocks producing temporary paralysis, and even death (Kitagawa 2000).

For the case of uniform current flow, the magnitude of the step voltage can be determined from Eq. (5.5) (see Fig. 5.3). Assuming a typical lightning current of 30 kA, a relatively low soil resistivity of 100 ohm-meters (Table 5.1), a distance between an individual's feet in the direction of current flow of 0.5 m (about 1.6 feet), and a distance from the lightning strike point to the individual of 10 m, the step voltage, from Eq. (5.5), is 2250 volts. If the resistance of a current path up one leg and down the other is assumed, for example, to be about 1000 ohms (the value will vary with type of shoes and degree of shoe and skin wetness), the current up one leg and down the other will reach about 2.3 amperes (2250 volts/1000 ohms). If the resistance of the current path through the legs is lower than 1000 ohms or if the soil resistivity is higher than 100 ohm-meters or if the lightning current is greater than 30 kA, the current through the legs will be greater than the 2.3 amperes calculated above. As noted in the previous section, four-legged animals are apparently often killed by step voltage between their front and hind legs.

Many cases have been reported in which groups of individuals were simultaneously shocked, injured, or killed. Carte et al. (2002) and Anderson (2001) report a case from South Africa in which 26 young girls, two adults, and seven dogs were sleeping in a 10 meter × 5 meter tent that was struck by lightning. Four girls and four dogs were killed. The two adults were unharmed, but almost all of the girls suffered some injuries including cataracts (8), burns (23), eardrum rupture (2), and skull fracture (2). Carte et al. (2002) discuss a number of other incidences of injury and death in relatively large groups including (a) 46 individuals at a concession stand in Ascot, England, of whom two died (Arden et al. 1956), and (b) 38 children playing soccer in England, of whom one died. Golde (1973) described a multiple fatal lightning incident on a mountain in Japan. The total group comprised 41 Japanese school children and five teachers who were surprised by a thunderstorm on an exposed mountain ridge. Eleven boys were killed. They all had lightning marks on their heads and necks. Seven were walking in a row with a distance from first to last of about 6 m. Fourteen other individuals sustained burn marks on their bodies and other injuries but survived. Anderson and Carte (1989) discuss the case of four golfers and their caddies, eight individuals in all, who were temporarily rendered unconscious by lightning, but subsequently and recovered. Fahmy *et al.* (1999) describe an event in which 17 individuals taking shelter under trees near a soccer field were injured by lightning. Seven children and four adults were admitted to the hospital. Ten sustained burns, four had suffered cardiopulmonary arrest, six had lower extremity paralysis, six had lost consciousness for up to one hour, and one suffered a ruptured eardrum. The only common feature among the ten burn victims was "small, circular, full-thickness burns involving the sides of the soles of the feet and the tips of the toes." These characteristic burns were on both feet of nine of the ten and were likely similar to those burns shown in Fig. 7.4. In most, but not all, group-strike cases the individuals in the group are fairly well separated from each other, and hence one or two may be subjected to a direct lightning strike or to a side flash or a surface arc from a directly struck individual or object, but the rest were probably victims of step voltages occurring between their feet, owing to a relatively uniform current flow in the Earth, or received severe shocks from unconnected upward leaders.

Since in this section we have estimated the probable currents that enter the human (or animal) body from direct strikes, unconnected upward leaders, and step voltages, it is natural to ask about the level of current that is sufficient to kill. It would appear that this question has not been satisfactorily answered, or perhaps there is not a satisfactory answer. The general effects of both 60 Hz power current and impulse current (in which category lightning falls) on humans and animals have been studied by Dalziel (1953, 1956) and Dalziel and Lee (1968). Ishikawa *et al.* (1985) have determined the level of impulse current necessary to kill a live rabbit. The proceedings of a symposium on the effects of electric shock have been published by Bridges (1985). The most mentioned mechanism of death is the induction of ventricular fibrillation, one of the several potentially fatal disturbances of cardiac rhythm. In the case of 60 Hz power frequency, if a human grips a current-carrying electrode with each hand, that individual will not be able to release his grip if the current exceeds about 0.01 amperes. At a current of 0.04 to 0.06 amperes asphyxia may occur from prolonged contraction of the chest muscles. For one-half cycle of the 60 Hz wave, a crude approximation to the lightning impulse, there exists about a 1 percent probability that cardiac fibrillation will occur in an adult human when the peak current is near 1 ampere. So 1 ampere of impulse current with a pulse width about 0.01 s may be near the threshold level for causing death. All of the investigators referenced above have concluded that the deleterious effects from impulse currents are best expressed in terms of the energy absorbed in the body (the action integral multiplied by body resistance; see Section 2.3) rather than in terms of the current. From Dalziel (1953), the threshold energy level for fibrillation of the 700 ohm individual in Fig. 7.9 is about 350 joules, consistent with an impulse current of a few amperes peak and a duration of 0.01 s. Ishikawa *et al.* (1985) found that the energy needed to assure death in rabbits was 63 joules per kilogram of body weight. If we linearly extrapolate to a 70 kilogram human, certainly a questionable extrapolation, the lethal energy is 4410 joules. So perhaps some

Fig. 7.10   An eighteenth-century personal lightning protection system proposed by Barbue-Dubourg. The air terminal is in the hat (bonnet) and the down conductor is grounded by contact with the Earth's surface. The etching is reproduced from the book by Figuier (1867).

hundreds of joules are the threshold for lethality and some thousands of joules delivered to the body are near-certainly lethal. If, for example, 7000 joules is lethal for our 700 ohm human, the action integral is 10 (7000 joules/700 ohms). This is equivalent to a body current of 1000 amperes flowing for $10\,\mu s$ (1000 amperes $\times$ 1000 amperes $\times$ 10 $\times$ 10^{-6} s), or a body current of 100 amperes flowing for 1 ms ($100\,\text{amperes} \times 100\,\text{amperes} \times 10^{-3}\,\text{s}$) or a body current of 10 amperes flowing for 0.1 s ($10\,\text{amperes} \times 10\,\text{amperes} \times 0.1\,\text{s}$). All of these different current levels are in the general ballpark of what unconnected upward leaders and direct lightning strikes can deliver to the body according to the model discussed earlier in this section and Fig. 7.9. Apparently, no specific physiological explanation of how the input energy affects the body has been advanced.

We have seen in the discussion above that the fact that most of the lightning current flashes over the body of a struck individual may well be a factor in preventing death and serious injury. The personal lightning protection system shown in Fig. 7.10, basically a conductor which extends from the hat down the body and drags along the Earth, operates to implement the same effect. The title of the illustration describes a hat (bonnet) air-terminal system for use by eighteenth-century Parisian women. The system was devised by Jacques Barbue-Dubourg and described by him in a letter dated 1773 to Benjamin Franklin, along with the description of an umbrella protection system with a lightning rod on the umbrella and a descending wire whose bottom dragged along the ground in the manner of the bonnet system (Willcox 1976). An etching of the umbrella lightning protection system is reproduced in the Frontispiece of the book you are reading. While both of these protection systems lack adequate grounding, they still might well save one's life, potential side flashes notwithstanding, or at least alter the otherwise existing electrical situation to the positive. The author has personally examined one case of an individual whose erected umbrella was struck by lightning, and who could well have been saved from death, or more serious injury than he did sustain, because the lightning current followed the path of the vertical umbrella mast before entering his body at the hip and thereby bypassed his body above the hip. While modes of lightning protection such as illustrated in Fig. 7.10 are certainly not to be recommended, one might well be better off with them than without them.

References

Arden, G. P., Harrison, S. H., Lister, J. and Maudsley, R. H. 1956. Lightning accident at Ascot. *Brit. Med. J.* **1**: 1450–1453.

Anderson, R. B. 2001. Does a fifth mechanism exist to explain lightning injuries? *IEEE Eng. Med. Biol.* Jan/Feb: 105–113.

Anderson, R. B. and Carte, A. E. 1989. Struck by lightning. *Archimedes* **31**(3): 25–29.

Anderson, R. B., Jandrell, I. R. and Nematswerani, H. E. 2002. The upward streamer mechanism versus step potentials as a cause of injuries from close lightning discharges. *Trans. S. Afr. Inst. Electr. Eng.* **93**(1): 33–37.

Andrews, C. J. 1992. Telephone-related lightning injury. *Med. J. Austr.* **157**: 823–826.

Andrews, C. J. and Darveniza, M. 1989. Telephone-mediated lightning injury: an Australian survey. *J. Trauma* **29**: 665–671.

Andrews, C. J., Cooper, M. A., Darveniza, M. and Mackerras, D. 1992. *Lightning Injuries: Electrical, Medical and Legal Aspects.* Boca Raton, FL: CRC Press.

Berger, K. 1971a. Blitzforschung und Personen-Blitzschutz. *ETZ (A)* **92**: 508–11.

Berger, K. 1971b. Zum Problem des Personenblitzschutzes. *Bull. Schweiz Elektrotech. Ver.* **62**: 397–99.

Blumenthal, R. 2005. Lightning fatalities on the South African highvelt. *Am. J. Forensic Med. Pathol.* **26**: 66–69.

Blumenthal, R. 2006. When thunder roars – go indoors. *S. Afr. Med. J.* **96**: 38–39.

Bridges, J. E., Ford, G. L., Sherman, I. A. and Vainberg, M. (eds). 1985. *Electric Shock Safety Criteria: Proc. 1st Int. Symp. Electric Shock Safety*. London: Pergamon Press.

Cannel, H. 1979. Struck by lightning – the effects upon men and the ships of HM Navy. *J. Roy. Nav. Med. Serv.* **65**: 165–170.

Carte, A. E., Anderson, R. B. and Cooper, M. A. 2002. A large group of children struck by lightning. *Ann. Emerg. Med.* **39**: 665–670.

Castren, J. A. and Kytila, J. 1963. Eye symptoms caused by lightning. *ACTA Opthalmol.* **41**: 139–143.

Cherington, M., Walker, J., Boyson, M. *et al.* 1999. Closing the gap on the actual numbers of lightning casualties and deaths. In *Proc. 11th Conf. Applied Climatology, Dallas, Texas*. American Meteorological Society, pp. 379–380.

Coates, L., Blong, R. and Siciliano, F. 1993. Lightning fatalities in Australia, 1824–1991. *Nat. Hazards* **8**: 217–233.

Cooper, M. A. 1980. Lightning injuries, prognostic signs for death. *Ann. Emerg. Med.* **9**: 134–138.

Cooper, M. A. 1995a. Emergent care of lightning and electrical injuries. *Semin. Neurol.* **15**: 268–278.

Cooper, M. A. 1995b. Myths, miracles, and mirages. *Semin. Neurol.* **15**: 358–361.

Cooper, M. A. and Andrews, C. J. 1995. Lightning injuries. In *Wilderness Medicine*, 3rd edn, ed. P. Auerbach. St. Louis: Mosby, pp. 261–289, (see also 4th edn, pp. 73–110).

Cooper, M. A. and Mashburn, S. 2005. Lightning strike and electric shock survivors. *Int. NeuroRehab.* **20**: 43–47.

Curran, E. B., Holle, R. L. and Lopez, R. E. 2000. Lightning casualties and damages in the United States from 1959 to 1994. *J. Clim.* **13**: 3448–3464.

Dalziel, C. F. 1953. A study of the hazards of impulse currents. *Power Apparatus Syst.* **72**, 1032–1040.

Dalziel, C. F. 1956. Effects of electric shock on man. *Trans. IRE Med. Electronics*, PGME-5, 44–62.

Dalziel, C. F. and Lee, W. R. 1968. Reevaluation of lethal electric currents. *IEEE Trans. Ind. Gen. Appl.* IGA-**4**, 467–476 & 676–677.

Elsom, D. M. 2001. Deaths and injuries caused by lightning in the United Kingdom: analyses of two databases. *Atmos. Res.* **56**: 325–334.

Fahmy, F. S., Brisden, M. D., Smith, J. and Frame, J. D. 1999. Lightning: the multisystem group injuries. *J. Trauma: Injury, Infection, and Critical Care* **46**: 937–940.

Figuier, L. 1867. *Les merveilles de la science, ou description populaire des inventions modernes*. Paris: Furne, Jouvet, pp. 569–597.

Golde, R. H. 1973. *Lightning Protection*. London: Edward Arnold.

Holle, R. L., López, R. E. and Navarro, B. C. 2005. Deaths, injuries, and damages from lightning in the United States in the 1890s in comparison with the 1990s. *J. Appl. Meteorol.* **44**: 1563–1573.

Holle, R. L., López, R. E. and Zimmerman, C. 1999. Updated recommendations for lightning safety. *Bull. Am. Meteorol. Soc.* **80**: 2035–2041.

Ishikawa, T., Ohashi, M., Kitigawa, N., Nagai, Y. and Miyazawa, T. 1985. Experimental study on the lethal threshold value of multiple successive voltage impulses to rabbits simulating multi-strike lightning flash. *Int. J. Biometeorol.* **29**(2), 157.

King, L. A. 1962. The voltage gradient of the free burning arc in air or nitrogen. In *Proc. 5th Int. Conf. Ionization Phenomena in Gases, Munich.* North Holland, pp. 871–877.

Kitagawa, N. 2000. The actual mechanisms of so-called step voltage injuries. *25th Int. Conf. Lightning Protection*, Rhodes, Greece, pp. 781–785.

López, R. E. and Holle, R. L. 1998. Changes in the number of lightning deaths in the United States during the twentieth century. *J. Clim.* **11**: 2070–2077.

López, R. E., Holle, R. L., Heitkamp, T. A. *et al.* 1993. The underreporting of lightning injuries and death in Colorado. *Bull. Am. Meteorol. Soc.* **74**: 2171–2178.

Pakiam, J. E., Choa, T. C. and Chia, J. 1981. Lightning fatalities in Singapore. *Meteorol. Mag.* **110**: 175–187.

Ten Duis, H. J. 1998. Lightning strikes: danger overhead. *Brit. J. Sport Med.* **32**: 276–278.

Walsh, K. M., Bennett, B., Cooper, M. A. *et al.* 2000. National athletic trainers' association position statement: lightning safety for athletics and recreation. *J. Athletic Training* **35**: 471–477.

Willcox, W. B. (ed.) 1976. The papers of Benjamin Franklin, Volume 20, January 1 through December 31, 1773. New Haven and London: Yale University Press.

Zimmerman, C., Cooper, M. A. and Holle, R. L. 2002. Lightning safety guidelines. *Ann. Emerg. Med.* **39**: 660–664.

8 Lightning warning

8.1   Overview

In the previous chapter we discussed the effects of lightning on humans and animals, and we considered those situations and activities that are safe and those that are unsafe in a thunderstorm. In Section 7.1 we noted that in any outdoor group activity, like the positioning and movement of the spectators at a professional golf tournament, one individual should be designated as the responsible "weather person," the primary person in charge of keeping track of the potential for lightning or other dangerous weather and of specifying when to get out of harm's way, using a plan that is already in place and tested. Only if one knows there is danger can appropriate action be taken to try to assure safety.

There are many situations in which it is not obvious that thunderstorms are approaching; for example, when the view of the storm and its lightning is obstructed or when nearby noise drowns out the sound of thunder. Nevertheless, the first line of defense in lightning warning is generally the visual observation of an approaching storm and the use of "flash to bang" thunder-ranging, that is, counting the time delay between seeing the light from the lightning and hearing its thunder. The time difference is about 5 seconds for each mile (about 3 seconds for each kilometer) of distance between the lightning and the observer, since sound travels about 1/5 mile (about 1/3 kilometer) per second while the light from a flash reaches the observer in a very small fraction of a second, that is, virtually instantaneously since light travels at 186 000 miles per second (300 000 kilometers per second). The popular and not unreasonable 30–30 rule (30 seconds–30 minutes) for lightning safety is this. The first 30: find a safe location at the first thunder-delay of 30 seconds, indicating that lightning is about 6 miles (about 10 kilometers) away; and the last 30: stay in the safe location for 30 minutes (the duration of a typical storm) following the last thunder heard from any distance.

As noted above, there are, unfortunately, many situations in which it is difficult (if not impossible) to hear thunder, particularly thunder from lightning 6 miles away and beyond. One location where it is important to hear thunder, but is often difficult to do so, is along the seashore where crashing ocean waves provide a loud (and esthetically pleasing) competing noise source. Another situation in which it is difficult to hear thunder is when operating a tractor, lawn mower, or similar motorized equipment, where the noise of the motor overwhelms all other sounds. Two major categories of lightning death and injury involve open water and farming (see Fig. 7.1), perhaps because thunder is often difficult to hear in those environments.

Besides watching for thunderclouds and lightning and listening for thunder, one can use information available from scientific instruments that detect storms and lightning. Data from both weather radars that detect precipitation (from which lightning activity can be indirectly inferred) and instruments that directly detect the occurrence of lightning and its location are made available to the general public on, for example, NOAA weather radio and cable TV's "The Weather Channel" in the United States. Individuals and organizations can purchase commercial services that will plot the location of each lightning strike point in the continental United States and Canada on a personal computer screen within about 30 seconds of its occurrence (e.g., https://thunderstorm.vaisala.com). Such a service is used, for example, by the University of Florida (UF) Athletic Association to provide light-ning warning for all organized outdoor athletic activities including practice sessions for football, baseball, softball, and soccer and for scheduled inter-University contests. Safe locations have been designated at each UF athletic venue for both athletes and spectators in the event of an approaching storm, although it is logistically impossible for all the fans at some large facilities, like the almost 100 000-seat UF football stadium, to move to safer places in the warning time available. Suggested solutions to this problem, including stadium protection via overhead ground wires (see Section 4.5), are discussed by Gratz and Noble (2006).

Lightning detection and locating systems generally measure and process the radio frequency (RF) signals that are characteristic of lightning discharges. The RF radiation detected by most of these systems is in the frequency range from a few kilohertz (kHz) to a few hundred kilohertz, the frequency range in which most of the electromagnetic energy of the lightning return stroke is radiated. In fact, audible static at the lower end of the AM band radio (300 kHz to 3 MHz) is a good indicator of lightning within about 50 km (about 30 miles). The louder the static, the closer the lightning, although accurate distance ranging is not possible on individual lightning flashes because large transient-static-producing events that are far away can produce similar levels of static as small events that are nearby, and different lightning flashes at any given distance can produce a wide range of RF signal amplitudes. There are many relatively simple, single-station devices that purport to locate lightning. Most operate like AM radios and use the amplitude of the radio static to gauge the distance to the individual lightning flashes, despite the inherent inaccuracy of this technique noted above. Some single-station devices use an optical detector and/or a magnetic direction finder in conjunction with an RF amplitude detector. In general, accurate lightning location is only possible if a number of spatially separated RF detectors are used in concert.

A single electromagnetic field sensor with sensitive electronics can detect the occurrence of lightning at distances as far away as several thousand kilometers. In fact, lightning signals at extremely low frequencies (3 Hz to 3 kHz) and the lower part of the very-low-frequency band (VLF: 3 kHz to 30 kHz) can circle the globe without too much attenuation. Lightning electromagnetic signals with frequencies greater than several kilohertz propagate to great distances in the so-called wave-guide mode, that is, by repeatedly bouncing off the conducting ionosphere (about 90 km above the Earth's surface) and the Earth, while those with lower frequencies (below the waveguide mode cutoff) are directly guided between the Earth and the ionosphere (in a mode termed quasi-transverse-electromagnetic). A network of about 20 long-range VLF lightning sensors currently covers the globe (e.g., Jacobson *et al.* 2006, Rodger *et al.* 2004, 2005), but the location accuracy of that network on individual flashes is limited, and only a very small fraction of the lightning flashes that occur is detected at such great distances. Long-range RF detection and location systems, as well as shorter-range RF systems, generally use loop antennas for measuring the lightning magnetic field, which, as we shall discuss in Section 8.2, provide information on the direction of the lightning, and vertical whip antennas (similar to an automobile radio antenna) or other elevated metallic antennas to detect the lightning electric field. The range to the lightning from very distant sensors is estimated by measuring the amplitudes and arrival times of different frequency components of the overall radiated electromagnetic signal since different frequency components propagate at different speeds and suffer different attenuations. The accuracy of such long-range estimates depends on a knowledge of the source characteristics (typical characteristics are reasonably well known but there is considerable variability from lightning to lightning) and of the physics of the lightning electromagnetic-wave interaction with a variable iono-sphere. If the data from many flashes striking in the same general area are accu-mulated at a distant detection station, one can "average" the data to determine a reasonably accurate location for the center of that group of flashes.

The most accurate RF lightning locating systems for determining the lightning ground strike-point can do so with an accuracy of better than 1 km over areas that are hundreds to thousands of kilometers in diameter. These systems necessarily use spatially separated multiple sensors that are precisely synchronized. In Section 8.2 we will discuss how individual RF sensors measuring various characteristics of the lightning electromagnetic radiation have been combined into systems that provide practical lightning locating and warning systems. One of the best examples of such a system is the North American Lightning Detection Network (NALDN), which consists of about 150 ground-based electric and magnetic field sensors that transmit their measured lightning data by satellite to a central station where the data are processed to determine the lightning strike locations over all of the continental United States and Canada. The NALDN sensors detect lightning electromag-netic signals in the frequency band from some tens of kilohertz to a few hundreds of kilohertz. The system is described and output data can be seen at https:// thunderstorm.vaisala.com. Systems similar to the NALDN are operating in more than 40 countries worldwide.

The quality of any lightning locating system is determined both by the accuracy of its locations and by the fraction of the actual lightning flashes that it detects, the so-called detection efficiency. For the NALDN, typical detection efficiency for flashes to ground is 85 to 90 percent. That is, 10 to 15 percent of the lightning is not located, mostly because the return strokes that constitute the flash are too small in current and radiated electric and magnetic fields to be detected. Return strokes with peak currents below 5 to 10 kA are usually not detected (Cummins *et al.* 1998, Jerauld *et al.* 2005). NALDN accuracy in determining the ground strike-point is stated to be about 500 m over all of North America; that is, half the lightning will be located with an accuracy better than 500 meters and half will be worse than 500 m. A map showing the ground flash density (the number of flashes to Earth per square kilometer per year) derived from the system is found in Fig. 1.5. To find ground flashes per square mile, multiply ground flashes per square kilometer by 2.6. More information on flash densities is found in Sections 1.2 and 1.4.

Lightning locating systems that operate in the very-high-frequency band (VHF: 30 to 300 MHz) reserved for TV and FM radio broadcasting, as opposed to locating systems like the NALDN which operate at frequencies below the AM radio band, are also discussed in Section 8.2. These VHF systems are not primarily intended to locate lightning ground strike-points but rather to image the whole lightning channel, both inside and outside the cloud, by locating the radiation sources of the myriad of small sparks that are involved in forming the channels of a lightning discharge.

The thunderstorm charging process produces a slowly varying (on a scale of seconds) electric field that can easily be detected at distances up to 10 or 20 kilometers. This field varies too slowly to be measured adequately with the RF sensors commonly used to detect lightning. Special electromechanical devices called "electric field mills" or just "field mills" are typically used to measure the cloud electric field, from which an estimate of the cloud charge location and magnitude can be derived. A drawing of a field mill is shown in Fig. 8.1. It operates as follows: an electric field antenna (the metal studs in Fig. 8.1) is alternately covered and uncovered (shielded and unshielded) by the mechanically rotated, grounded metal plate above the antenna, so that the fraction of the nearly static cloud electric field that reaches the antenna (and induces a surface charge on the antenna where that field terminates) varies periodically from all to none. Thus the cloud electric field is changed from an essentially dc signal to a time-varying signal by the shielding and unshielding of the antenna. The electronics then processes the time-varying signal (in Fig. 8.1, the current flowing through the resistor R due to the time-varying charge induced on the metal studs), a relatively easy task compared with processing a nearly static signal. Because of the time it takes to mechanically shield and unshield the antenna, the time-resolution of field mills is generally limited to a fraction of a second, at best about a millisecond. In addition to detecting the presence of thundercloud charge, field mills can record the changes in the cloud electric field due to lightning and thus provide information on the lightning charge source location and magnitude. The NASA Kennedy Space Center (KSC) operates a network of 20 to 30 field mills for the purpose of detecting thundercloud and lightning charges in the vicinity of KSC (e.g., Jacobson and Krider 1976), information that is used to ensure safety in all civilian and military spacecraft launches at

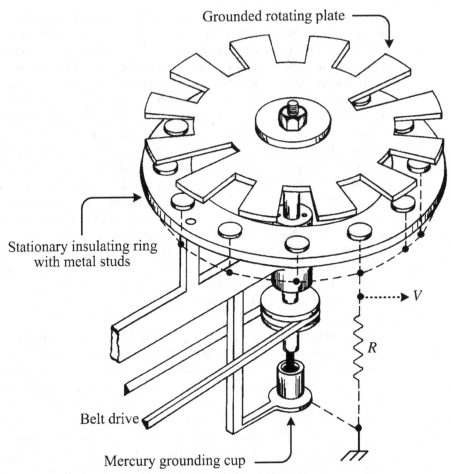

Grounded rotating plate

Stationary insulating ring
with metal studs

V

R

Belt drive

Mercury grounding cup

Fig. 8.1   A drawing of a field mill, one of the original designs. Adapted from Malan (1963).

KSC and the adjacent military launch facilities. Single field mill sensors are inherently much less accurate in determining the presence of cloud charges and charge variations than are multiple field mills separated by some kilometers and analyzed in concert.

Optical "lightning-mappers" resident on Earth-orbiting satellites detect the light produced by lightning after that light is scattered by the cloud surrounding the lightning source. Hence, such satellite-based sensors can locate the lightning in latitude and longitude to an accuracy of about the diameter of a typical cloud, 10 km or so. Satellite-based sensors have difficulty distinguishing between cloud discharges (including intracloud, intercloud, and cloud-to-air, see Section 1.1) and cloud-to-ground discharges. All lightning-sensing satellites that have operated or are operating at the time of this book's writing have been launched into low Earth orbit where they can view only a relatively small part of the Earth's surface at any instant. Undoubtedly, it is just a matter of time until all lightning on Earth is charted from satellites that view the entire Earth from high and synchronous Earth orbit (sychronous meaning stationary relative to the Earth). More details about the detection of lightning from satellites are found in Section 8.3.

8.2 RF location techniques

Two main techniques are used in ground-based RF lightning locating systems: magnetic direction finding (MDF) and pulse time-of-arrival (TOA). The NALDN noted in Section 8.1 uses both techniques in combination. We will now examine these techniques separately.

8.2.1 The magnetic-field direction finding technique

Two vertically oriented, perpendicular loops of wire, with the individual loop planes oriented north/south and east/west (as shown in Fig. 8.2 and Fig. 8.3) can be used to determine the direction to a vertical current source. Similar direction finding systems on aircraft, measuring the directions to fixed ground-based RF transmitters, have been the workhorse of aircraft navigation for most of the history of aviation, until the advent of the Global Positioning System (GPS) of satellites. The output voltage from a given loop of wire is proportional to the component of the magnetic field that is perpendicular to the plane of the loop via Faraday's Law (see Section 2.3, Fig. 2.7). For a vertical current source, like the lower part of a cloud-to-ground lightning channel, the magnetic field forms concentric circles around the source, as illustrated in Fig. 8.2, Fig. 8.3, and Fig. 2.7. For example, when the plane of the loop is oriented east/west, it receives a maximum signal if the source is east or west of the antenna, while the orthogonal (north/south) loop receives no signal, a situation that is illustrated in Fig. 8.2. As shown in Fig. 8.3, the signal in the NS loop varies as the cosine (a trigonometric function) of the angle θ between north and the source as viewed from the antenna, while the signal in the EW loop varies as the sine (another trig function) of the same angle. It follows that the ratio of the EW to NS signals is proportional to the tangent (the sine divided by the cosine) of the angle between north and the source as viewed from the antenna, and hence, as long as the current source radiating the magnetic field is vertical, the direction (angle) to the source can be determined by measuring the ratio of the voltages detected by the two perpendicular loops.

Before the development of weather radar in the 1940s, magnetic direction finding on lightning was the primary means of identifying and mapping thunderstorms. That is, the lightning signals were used to infer the presence of a thunderstorm at the location of the lightning. Now it is common to use weather radar echoes greater than a given strength to infer that lightning is occurring; that is, to indicate there is heavy enough precipitation in the cloud to produce the electrical charge separation that produces lightning.

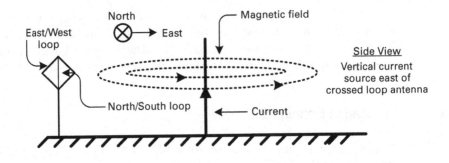

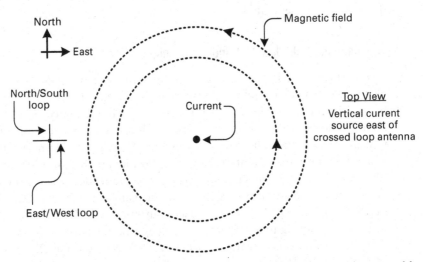

Fig. 8.2   The magnetic field of a vertical current source and its measurement with a crossed-loop magnetic-field direction finder.

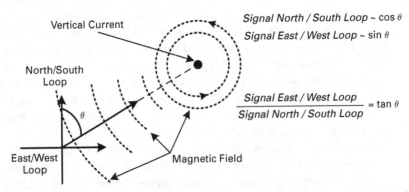

Fig. 8.3   Determining the azimuth angle to a vertical current source with magnetic direction finding.

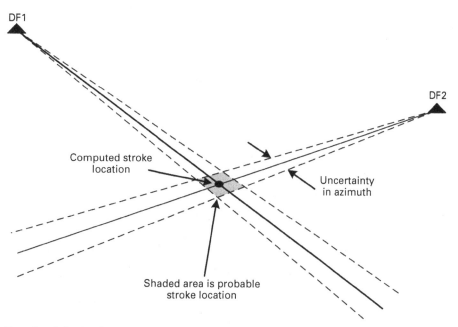

DF1

DF2

Computed stroke
location

Uncertainty
in azimuth

Shaded area is probable
stroke location

Fig. 8.4   Locating lightning from the intersection of two magnetic direction finding vectors and
the errors involved.

In the 1920s, Watson-Watt and Herd (1926) developed a crossed-loop
magnetic direction finder for lightning using a pair of orthogonal loop antennas
tuned to a frequency near 10 kHz, roughly the dominant electromagnetic frequency radiated by the individual return strokes in flashes to ground. This
instrument is referred to as a narrowband magnetic direction finder to distinguish it from later direction finders that detected a wider range of frequencies.
The azimuth angle θ (measured clockwise from north as in Fig. 8.3) to the
discharge was obtained by displaying the north–south and east–west loop outputs simultaneously on the two perpendicular axes of an oscilloscope screen,
such that the resulting line on the screen pointed in the direction to the discharge.
Two such direction finders at different locations were sufficient to plot the
position of a discharge from the intersection of the simultaneous direction
vectors, as illustrated in Fig. 8.4 where the uncertainty (or error) in the direction
is indicated by the dotted lines around the two measurements (solid lines).
Similar lightning locating systems were used in many countries before and during
World War II. For example, during World War II the British Meteorological
Office operated a narrowband (a 250 Hz bandwidth centered at 9 kHz) magnetic
direction-finding network containing seven sensors, located both in the United
Kingdom and in the Mediterranean region, in support of the activities of the
aircraft of the Royal Air Force (WMO 1955). The British magnetic direction-
finding network could identify thunderstorms occurring in an area ranging from
the United Kingdom to North Africa.

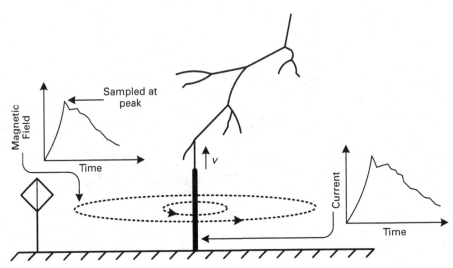

Fig. 8.5   The current in the developing return stroke channel and the radiated magnetic field during the first few microseconds of the return stroke. The return stroke propagates upward at speed *v* along a reasonably straight and vertical channel bottom.

A major disadvantage of narrowband magnetic direction finders is that when lightning occurs at distances less than about 200 km, those sensors exhibit an inherent azimuth angle error of the order of 10 degrees (Nishino *et al.* 1973, Kidder 1973). These relatively large errors are caused by the detection of magnetic field components from non-vertical channel sections, including branches and in-cloud channels, and by ionospheric reflections of the radiated magnetic field signal, so-called skywaves, whose magnetic field directions are improperly oriented for direction finding to the ground strike-point. In general, the so-called polarization errors become less prominent and the azimuth errors associated with them become smaller as the distance to the lightning increases beyond about 200 km.

To overcome the problem of relatively large polarization errors and the associated large position errors inherent in the narrowband magnetic direction finder approach to locating lightning within about 200 km, wideband magnetic field sensors that can sample a given instant of the time-varying magnetic field signal were developed in the early 1970s as part of research at the University of Arizona and the University of Florida. So-called gated, wideband direction finders operate by sampling the NS and EW components of the return stroke magnetic field just at the initial peak (corresponding to the initial peak of the current shown in Fig. 1.8, Fig. 1.9, Fig. 2.1, and Fig. 2.2), so that just the radiation from the bottom portion of the channel, during the first microseconds of the return stroke's upward propagation, is sampled, as illustrated in Fig. 8.5. Since the bottom of the channel tends to be more or less straight and vertical (but not exactly so), the magnetic field is nearly horizontally polarized at those early times in the return stroke waveform; that is, it forms horizontal circles surrounding the source. Additionally, a gated, wideband magnetic direction finder does not record ionospheric reflections since

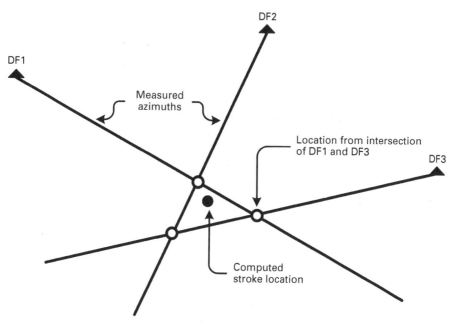

Fig. 8.6   Locating lightning from the intersection of three direction finding vectors.

those reflections arrive at the sensor after the initial peak magnetic field has been
sampled. The operating bandwidth of such sensors is typically from a few kilohertz
to a few hundred kilohertz, frequencies below most of the AM radio band in order
to avoid interference from that source. The original gated, wideband instrument is
described by Krider *et al.* (1976) along with a determination of its angle error, about
one degree, that was measured by comparing direction-finder results with accurate
video recordings of the causative return strokes. Further, Krider *et al.* (1980) have
described a gated, wideband magnetic direction finder that responds primarily to
the return strokes in ground flashes. The magnetic field waveforms of return
strokes are electronically separated from the waveforms of in-cloud processes and
various non-lightning sources by taking advantage of the unique characteristics of
the shape of the return stroke magnetic field waveform.

  As shown in Fig. 8.4, the intersection of two simultaneous direction (azimuth)
vectors measured at different locations, lines from the sensor to the apparent
source, can be used to determine a stroke location. However, that location will
contain errors because each azimuth vector has some random angular error and
may also have some fixed (systematic) angular error, the latter generally caused
by electrically conducting objects such as structures or power lines being located
too near the antenna. If a three-sensor network is used, each pair of sensors yields
a location, with the distance between the three locations providing some measure
of the overall location error, as illustrated in Fig. 8.6. If three or more sensors
measure the azimuth to a given return stroke and if the random errors in the
measurements are known, an optimal estimate of the actual location, the so-called

most probable location, can be found using a least-squares minimization-of-error technique which also provides an estimate of the error in the location (e.g., Hiscox *et al.* 1984, Koshak *et al.* 2004). In fact, an elliptical area (a "confidence ellipse") can be drawn within which there is a given probability, say 99 percent, that lightning did occur, with the most probable location of the strike-point being at the center of that area.

The US National Lightning Detection Network (NLDN) began commercial operation in 1989. It originated from a combination of several government, industry, and university networks that were operated separately, starting in the late 1970s. The original NLDN sensors were gated, wideband magnetic direction finders. At present, an improved and expanded network, the North American Lightning Detection Network (NALDN) noted in Section 8.1, uses a combination of the gated, wideband magnetic direction finding technique and a gated, wideband time-of-arrival technique (Cummins *et al.* 1998). We discuss next the general features of lightning locating using the time-of-arrival technique.

8.2.2   The time-of-arrival technique

A single time-of-arrival (TOA) sensor measures the time at which a pre-determined portion of the lightning electromagnetic field arrives at the sensing antenna. The time that the signal was radiated, t_0 in Fig. 8.7, is unknown. With two TOA sensors, one can measure the difference in the arrival times between the two sensors. For a given time difference, the source must be located somewhere on a hyperbola

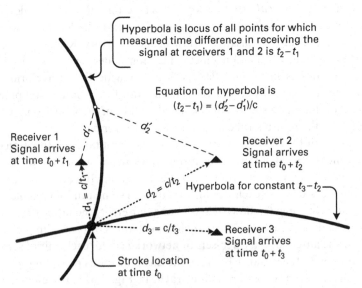

Fig. 8.7   Locating lightning using the difference in time-of-arrival technique. The speed of light is designated c.

passing between the two stations, as illustrated in Fig. 8.7. The hyperbola represents the locus of all possible points of source origin that could produce the given time difference. For example, the hyperbola representing the time difference $t_2 - t_1$ passes through the actual stroke location and is symmetrical about a line from receiver 1 and receiver 2. If the source is very far away compared with the distance between the stations, the distant portion of the hyperbola is essentially a straight line, a direction-finding vector.

For a TOA system, the outputs of four TOA sensors (three time differences) are generally needed to calculate a unique location. This is the case because, if the stroke is in an area outside the receiving stations, two time differences will produce two hyperbolas that intersect at two locations. One is the actual stroke location, and one will be erroneous. The stroke illustrated in Fig. 8.7 is sufficiently inside the area of the receiving stations that the two hyperbolas representing the locus of all points for which $t_2 - t_1$ is a constant value and $t_3 - t_2$ is a constant value, respectively, intersect at one point only. Since the actual location cannot be known *a priori*, inside or outside the network, four sensors yielding three time differences are necessary. The TOA technique of determining the lightning location from the intersection of three hyperbolas via three time differences is not the only technique that can determine location from timing measurements. If the absolute time-of-arrival can be measured with a small error at a number of stations, as is possible using GPS timing systems, a least-squares minimization-of-error technique can be used to find the most probable time that the lightning occurred and, with that information, its location. This technique is used in the present NALDN which, as noted above, combines the magnetic direction-finder and TOA techniques for maximum accuracy.

The first TOA network that detected lightning over a relative large area is described by Lewis *et al.* (1960). The system was composed of four stations over 100 km apart located in the northeastern United States and received signals in a bandwidth between 4 kHz and 45 kHz. Lewis *et al.* (1960) used the differences in the time of arrival of distant waveforms at two of the receiving stations to determine directions to the causative lightning discharge in western Europe. As noted above, the distant portion of the hyperbola representing a measured time difference is essentially a straight line, a direction-finding vector. The resultant directions compared favorably with the locations of lightning flashes reported by the British Meteorological Office's narrowband magnetic direction-finding network discussed above. Interestingly, the British Meteorological Office's MDF network, used extensively to locate lightning and thunderstorms during and after World War II, was converted in the 1980s to a time-of-arrival system (Lee 1986a,b, 1989a,b, 1990). The present British system has seven stations with separations between the stations ranging from about 250 to 3300 km. It operates in the 2 kHz to 18 kHz frequency range. The sensors are located in the United Kingdom, Gibraltar, and Cyprus. The system's stated flash location error is 2 to 20 km, and thus the system is useful primarily for detecting storm areas. No attempt is made to distinguish between cloud and ground flashes.

The first commercial TOA system, called the Lightning Positioning and Tracking System (LPATS), was introduced in the 1980s (e.g., Lyons *et al.* 1989). The LPATS, operating at similar frequencies to the gated, wideband MDF system described above, used electric field whip antennas at four or more stations separated by 200 to 400 km to determine locations via the measured differences between signal arrival times at each station. In the frequency band used, return stroke waveforms are generally the largest and hence most easily identified. Early versions of LPATS were synchronized by measuring the standard timing signals from LORAN-C or other Earth-based timing systems (LORAN-C is a long-range radio-navigation system that serves the continental United States as well as parts of Alaska and coastal waters) at each of the individual stations, while later versions used GPS to synchronize station clocks. A recent wideband TOA system, used primarily for research, has been developed by the Los Alamos National Laboratory (Shao *et al.* 2006).

The wideband TOA and MDF systems discussed above detect and process lightning signals at frequencies typically below some hundreds of kilohertz. These systems are primarily intended to locate the ground strike-point of individual strokes in lightning flashes to ground, although some level of detection of cloud discharges is also possible. Another type of TOA system operates in the VHF band (30 to 300 MHz) with center frequencies in the tens to hundreds of megahertz range and bandwidths between 5 and 10 MHz. These very high frequency systems can provide a radio frequency image of the whole lightning channel. Propagating leaders, which originate in the cloud charge and travel throughout the cloud and often to ground, emit pulses of RF energy in the bandwidth of these systems during the process of leader extension (stepping). It is thus possible to track the leader tip as a function of time and hence its overall path using the TOA technique. The most advanced system of this type is the New Mexico Tech Lightning Mapping Array (LMA) (e.g., Rison *et al.* 1999, Thomas *et al.* 2004) which is finding widespread use in both research and in operational situations such as bad-weather aircraft traffic control around major metropolitan areas. The LMA can locate the sources of impulsive RF radiation in three spatial dimensions and time with a spatial uncertainty of tens of meters. It does so by measuring the arrival times of RF events to an accuracy of tens of nanoseconds at a network of over 10 ground-based stations covering an area of typically 60 km in diameter. The impulsive RF signals are detected in an unused part of the television band, usually at 60–66 MHz. Thomas *et al.* (2004) review the history of VHF TOA systems for lightning location. The pioneering system was developed and used very successfully for research by Proctor in South Africa (e.g., Proctor 1971, 1981; Proctor *et al.* 1988). The system had five stations and received radiation of 300 MHz with a 5 MHz bandwidth. Building on Proctor's work, the Lightning Detection and Ranging (LDAR) system was developed in the 1970s by Lennon at the NASA Kennedy Space Center (KSC), Florida for operational use at KSC and the adjacent Cape Canaveral Air Force Station. An advanced version of LDAR is now in use there as an aid to assessing spacecraft launch conditions (e.g., Boccippio *et al.* 2001).

8.3   Detection from satellites

With the advent of Earth-orbiting satellites, it has become possible, in principle, to measure the worldwide lightning activity by detecting the light or the radio frequency signals emitted in the upward direction by both cloud and cloud-to-ground discharges. During the 1990s NASA researchers developed a satellite lightning mapper designed for geostationary orbit, but such a sensor is still not in place (Davis *et al.* 1983, Christian *et al.* 1989). This sensor is a CCD (charge-coupled device) optical array with electronics capable of detecting transient luminosity from lightning, even during the day time. It is designed to detect lightning from geostationary altitudes with a spatial resolution of 10 km and with a temporal resolution of 1 ms. This sensor has been combined with lenses that will provide coverage over much of North America, including all of the contiguous United States and nearby ocean area, Central America, South America, and the inter-tropical convergence zone. The system is designed to detect 90 percent of the flashes that occur. Such an optical system will allow mapping of both cloud and ground lightning continuously on continental scales.

A number of satellites have been launched that include optical sensors that detect lightning activity, some as their primary objective, others pointed at the Earth for various other reasons, mostly detecting nuclear tests and missile launches. The satellites used to date have recorded only a small fraction of the lightning flashes because those satellites were in relatively low orbit and hence spent a relatively short time over any given storm, as noted earlier, along with other limitations. Nevertheless, from orbiting satellites it has been possible to estimate local and worldwide flash densities as a function of season. These data are of particular interest in regions of the world that have no other means of lightning detecting and locating, such as in portions of Africa and South America.

Optical detectors derived from the design of the geostationary mapper referred to above have been flown on two satellites in low Earth orbit. The Optical Transient Detector (OTD) was launched on the Microlab-1 (recently renamed OV-1) satellite in 1995 into an Earth orbit of 735 km altitude with an inclination of 70 degrees with respect to the Equator, a near-polar orbit (Christian *et al.* 1992, 1996, 2003). The OTD operated for five years and stopped sending data in April, 2000. It had a 100 degree field of view and hence observed a 1300 km × 1300 km region, about 1/300 of the Earth's surface at any instant, orbiting the Earth in 100 minutes with a nominal spatial resolution of about 10 km and a nominal time resolution of 2 ms. From a comparison of OTD and National Lightning Detection Network data, it was determined that the OTD detection efficiency for ground flashes was about 45 to 70 percent, and that it was likely to be slightly higher for cloud flashes (Boccippio *et al.* 2000). Because of its orbit, the OTD never observed a given location for more than a few minutes per day. Data from the OTD are found on the website http:// thunder.msfc.nasa.gov/otd.html. Flash density maps for July/August 1995 and January/February 1996 are given by Christian and Latham (1998).

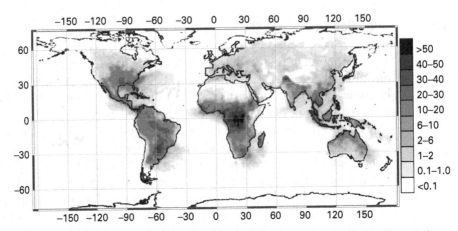

Fig. 8.8   A global flash density map determined from the OTD and LIS satellite data. The scale for the flash density in flashes per square kilometer per year is given on the right, varying from over 50 to less than one. Courtesy of H. J. Christian, NASA Marshall Space Flight Center.

The second orbital lightning mapper is called the Lightning Imaging Sensor (LIS). It was launched aboard the Tropical Rainfall Measuring Mission (TRMM) Observatory in 1997, and it views a 600 km × 600 km region, a given point on the Earth being observed for almost 90 s as the TRMM satellite circles the Earth at 7 km s^{-1}. The TRMM Observatory orbit has an inclination of 35 degrees so that LIS can observe lightning between (atitudes of 35 degrees south and 35 degrees north. Its estimated flash detection efficiency is near 90 percent. Data from the LIS are found on the website http://thunder.msfc.nasa.gov/lis.html.

A map of global lightning flash density based on data from the two satellite detectors discussed above, five years of OTD data and three years of LIS data, is shown in Fig. 8.8.

The Japanese Ionospheric Sounding Satellite ISS-b recorded RF radiation from lightning at 2.5, 5, 10, and 25 MHz, providing a two-year worldwide lightning map derived from RF data (Kotaki *et al.* 1981a,b, Kotaki and Katoh 1983). Lightning RF emissions in the VHF band (30 to 300 MHz) have also been observed by the Blackbeard receiver aboard the Alexis satellite (Zuelsdorf *et al.* 1997, 1998a,b, 2000). Blackbeard observed almost exclusively high energy, narrow pulse VHF emissions that occurred in pairs and came to be known as "transionospheric pulse pairs" or TIPPs. TIPPs are apparently generated by small, intense cloud discharges, often referred to as "compact intracloud discharges," producing narrow (about 10 μs) bipolar pulses of electric and magnetic fields that are also observed on the ground. The first pulse of the pair observed on the satellite is thought to travel directly from the in-cloud discharge to the satellite and the second pulse is thought to be caused by a reflection of the initial radiation off the ground and then upward and through the ionosphere to the satellite receiver. In part to explore further the origin of the TIPPs, the FORTE satellite, containing both RF and optical sensors, was launched in 1997 (Jacobson *et al.* 1999, 2000). Correlated lightning optical and RF signals have been studied to show that, as viewed from Earth orbit, the detected light from lightning, even for return strokes to ground, primarily emanates from within the cloud (Suszcynsky *et al.* 2000).

References

Boccippio, D. J., Koshak, W., Blakeslee, R. *et al.* 2000. The Optical Transient Detector (OTD): instrument characteristics and cross-sensor validation. *J. Atmos. Ocean. Technol.* **17**: 441–458.

Boccippio, D. J., Heckman, S. and Goodman, S. J. 2001. A diagnostic analysis of the Kennedy Space Center LDAR network: 1. Data characteristics. *J. Geophys. Res.* **106**: 4769–4786.

Christian, H. J. and Latham, J. 1998. Satellite measurements of global lightning. *Q. J. Roy. Meteorol. Soc.* **124**: 1771–1773.

Christian, H. J., Blakeslee, R. J. and Goodman, S. J. 1989. The detection of lightning from geostationary orbit. *J. Geophys. Res.* **94**: 13329–13337.

Christian, H. J., Blakeslee, R. J. and Goodman, S. J. 1992. *Lightning Imaging Sensor for the Earth Observing System.* NASA Tech. Memorandum 4350.

Christian, H. J., Driscoll, K. T., Goodman, S. J. *et al.* 1996. The Optical Transient Detector (OTD). In *Proc. 10th Int. Conf. Atmospheric Electricity, Osaka, Japan*, pp. 368–371.

Christian, H. J., Blakeslee, R. J., Boccippio, D. J. *et al.* 2003. Global frequency and distribution of lightning as observed from space by the Optical Transient Detector, *J. Geophys. Res.* **108**: 4005, doi:10.1029/2002JD002347.

Cummins, K. L., Murphy, M. J., Bardo, E. A. *et al.* 1998. A combined TOA/MDF technology upgrade of the U.S. National Lightning Detection Network. *J. Geophys. Res.* **103**: 9035–9044.

Davis, M. H., Brook, M., Christian, H. *et al.* 1983. Some scientific objectives of a satellite-borne lightning mapper. *Bull. Am. Meteorol. Soc.* **64**: 114–119.

Gratz, J. and Noble, E. 2006. Lightning safety and large stadiums. *Bull. Am. Meteorol. Soc.* **87**, No.9, 1187–1194.

Hiscox, W. L., Krider, E. P., Pifer, A. E., and Uman, M. A. 1984. A systematic method for identifying and correcting "site errors" in a network of magnetic direction finders. Preprint, *Proc. Int. Aerospace and Ground Conf. Lightning and Static Electricity, Orlando, Florida*, pp. 7-1–7-5, National Interagency Coordination Group.

Jacobson, E. A. and Krider, E. P. 1976. Electrostatic field changes produced by Florida lightning, *J. Atmos. Sci.* **33**: 113–117.

Jacobson, A. R., Knox, S. O., Franz, R. and Enemark, D. C. 1999. FORTE observations of lightning radio-frequency signatures: capabilities and basic results. *Radio Sci.* **34**: 337–354.

Jacobson, A. R., Cummins, K. L., Carter, M. *et al.* 2000. FORTE radio-frequency observations of lightning strokes detected by the National Lightning Detection Network. *J. Geophys. Res.* **105**: 15653–15662.

Jacobson, A. R., Holzworth, R. Harlin, J., Dowden, R. and Lay, E. 2006. Performance assessment of the world wide lightning location network (WWLLN), using the Los Alamos sferic array (LASA) as ground truth. *J. Atmos. Ocean. Technol.* **23**:1082–1092.

Jerauld, J., Rakov, V. A. Uman, M. A., Rambo, K. J. and Jordan, D. M. 2005. An evaluation of the performance characteristics of the U.S. national lightning detection network in

Florida using rocket-triggered lightning. *J. Geophys. Res.* **110**, D19106, doi:10.1029/ 2005JD005924.

Kidder, R. E. 1973. The location of lightning flashes at ranges less than 100 km. *J. Atmos. Terr. Phys.* **35**: 283–290.

Koshak, W. J., Solakiewicz, R. J., Blakeslee, R. J. *et al.* 2004. North Alabama Lightning Mapping Array (LMA): VHF source retrieval algorithm and error analyses. *J. Atmos. Ocean. Technol.* **21**: 543–558.

Kotaki, M. and Katoh, C. 1983. The global distribution of thunderstorm activity observed by the ionospheric sounding satellite (ISS-B). *J. Atmos. Terr. Phys.* **45**: 833–847.

Kotaki, M., Kuriki, I., Katoh, C. and Sugiuchi, H. 1981a. Global distribution of thunderstorm activity observed with ISS-b. *J. Radio Res. Lab. Tokyo* **28**: 49–71.

Kotaki, M., Sugiuchi, H. and Katoh, C. 1981b. *World Distribution of Thunderstorm Activity Obtained from Ionosphere Sounding Satellite-b Observations June 1978 to May 1980.* Japan: Radio Research Laboratories, Ministry of Posts and Telecommunications.

Krider, E. P., Noggle, R. C. and Uman, M. A. 1976. A gated wideband magnetic direction finder for lightning return strokes. *J. Appl. Meteorol.* **15**: 301–306.

Krider, E. P., Noggle, R. C., Pifer, A. E. and Vance, D. L. 1980. Lightning direction-finding systems for forest fire detection. *Bull. Am. Meteorol. Soc.* **61**: 980–986.

Lee, A. C. L. 1986a. An experimental study of the remote location of lightning flashes using a VLF arrival time difference technique. *Q. J. Roy. Meteorol. Soc.* **112**: 203–229.

Lee, A. C. L. 1986b. An operational system for the remote location of lightning flashes using a VLF arrival time difference technique. *J. Atmos. Ocean. Technol.* **3**: 630–642.

Lee, A. C. L. 1989a. The limiting accuracy of long wavelength lightning flash location. *J. Atmos. Ocean. Technol.* **6**: 43–49.

Lee, A. C. L. 1989b. Ground truth confirmation and theoretical limits of an experimental VLF arrival time difference lightning flash locating system. *Q. J. Roy. Meteorol. Soc.* **115**: 1146–1166.

Lee, A. C. L. 1990. Bias elimination and scatter in lightning location by the VLF arrival time difference technique. *J. Atmos. Ocean. Technol.* **7**: 719–733.

Lewis, E. A., Harvey, R. B. and Rasmussen, J. E. 1960. Hyperbolic direction finding with sferics of transatlantic origin. *J. Geophys. Res.* **65**: 1879–1905.

Lyons, W. A., Moon, D. A., Schuh, J. A., Pettit, N. J. and Eastman, J. R. 1989. The design and operation of a national lightning detection network using time-of-arrival technology. In *Proc. 1989 Int. Conf. Lightning and Static Electricity, Bath, England*, pp. 2B.2.1–8.

Malan, D. J. 1963. *Physics of Lightning.* London: The English Universities Press Ltd.

Nishino, M., Iwai, A. and Kashiwagi, M. 1973. Location of the sources of atmospherics in and around Japan. In *Proc. Res. Inst. Atmospherics, Nagoya Univ. Japan* **20**: 9–21.

Proctor, D. E. 1971. A hyperbolic system for obtaining VHF radio pictures of lightning. *J. Geophys. Res.* **76**: 1478–1489.

Proctor, D. E. 1981. VHF radio pictures of cloud flashes. *J. Geophys. Res.* **86**: 4041–4071.

Proctor, D. E., Uytenbogaardt, R. and Meredith, B. M. 1988. VHF radio pictures of lightning flashes to ground. *J. Geophys. Res.* **93**: 12683–12727.

Rison, W., Thomas, R. J., Krehbiel, P. R., Hamlin, T. and Harlin, J. 1999. A GPS-based three-dimensional lightning mapping system: initial observations in central New Mexico. *Geophys. Res. Lett.* **26**: 3573–3576.

Rodger, C. J., Brundell, J. B., Dowden, R. L. and Thomson, N. R. 2004. Location accuracy of long distance VLF lightning location network. *Ann. Geophys.* **22**: 747–758.

Rodger, C. J., Brundell, J. B., Dowden, R. L. and Thomson, N. R. 2005. Location accuracy of VLF World Wide Lightning Location (WWLL) network: post-algorithm upgrade. *Ann. Geophys.* **23**: 277–290.

Shao, X.-M., Stanley, M., Regan, A. *et al.* 2006. Total lightning observations with the new and improved Los Alamos sferic array (LASA). *J. Atmos. Ocean. Technol.* **23**: 1273–1288.

Suszcynsky, D. M., Kirkland, M. W., Jacobson, A. R. *et al.* 2000. FORTE observations of simultaneous VHF and optical emissions from lightning: basic phenomenology. *J. Geophys. Res.* **105**: 2191–2201.

Thomas, R. J., Krehbiel, P. R., Rison, W. *et al.* 2004. Accuracy of the lightning mapping array. *J. Geophys. Res.* **109**, D14207, doi:10.1029/2004JD004549.

Watson-Watt, R. A. and Herd, J. F. 1926. An instantaneous direct-reading radio goniometer. *J. Inst. Electr. Eng.* **64**: 611–622.

World Meteorological Organization (WMO). 1955. *Technical Note 12, Atmospheric Techniques*. Geneva: Secretariat of the World Meteorological Organization.

Zuelsdorf, R. S., Strangeway, R. J., Russel, C. T. *et al.* 1997. Trans-ionospheric pulse pairs (TIPPs): their geographic distribution and seasonal variations. *Geophys. Res. Lett.* **24**: 3165–3168.

Zuelsdorf, R. S., Casler, C., Strangeway, R. J. and Russel, C. T. 1998a. Ground detection of trans-ionospheric pulse pairs by stations in the National Lightning Detection Network. *Geophys. Res. Lett.* **25**: 481–484.

Zuelsdorf, R. S., Strangeway, R. J., Russell, C. T. and Franz, R. 1998b. Trans-ionospheric pulse pairs (TIPPs): their occurrence rates and diurnal variation. *Geophys. Res. Lett.* **25**: 3709–3712.

Zuelsdorf, R. S., Franz, R. C., Strangeway, R. J. and Russell, C. T. 2000. Determining the source of strong LF/VLF TIPP events: implications for association with NPBPs and NNBPs. *J. Geophys. Res.* **105**: 20725–20736.

9 Airships, airplanes, and launch vehicles

9.1   Overview

The metal skin of a modern airplane can be considered a good approximation both to a Faraday cage and to the outer surface of a topological shielded system (see Section 3.1). As such, the skin provides the primary lightning protection for the aircraft. Generally, when lightning strikes a metal airplane, the lightning current remains in the skin of the plane as it flows between entrance and exit points. If the shielding by the plane's skin were perfect, there would be no danger to the interior electronics or to the fuel in the wings of the airplane. Unfortunately, there are openings (apertures) such as windows in the metal skin and antennas that project through insulated areas in the skin, both of which may serve as entry points for lightning electromagnetic fields. Additionally, the plane's aluminum skin is not always thick enough to avoid direct damage by a severe lightning charge flowing into the skin (see Section 2.3). Lightning often disables interior aircraft electronics, as we shall discuss in Section 9.2, and occasionally lightning can burn through the aircraft's skin, igniting fuel or releasing hydraulic fluids, examples of which are given in Section 9.3. In Section 9.4, we will briefly consider the standards for testing aircraft to make more certain they can withstand a lightning strike without serious consequences.

Contrary to the common view, most lightning discharges that strike airplanes in flight are initiated by the planes themselves. The lightning would not have occurred if the plane had not been present. Although long suspected, the fact that an airplane can "trigger" its own lightning was first demonstrated convincingly in the 1980s (Mazur *et al.* 1984). Even in the case of an airplane struck by a lightning flash that has been initiated independent of the plane, an event that represents perhaps 10 percent of all strikes to planes, there is likely a significant electrical discharge initiated from the plane toward the incoming lightning leader. A video frame showing the initiation of lightning by a commercial aircraft, just after takeoff from an airport in Japan and still at relatively low altitude, is found in Fig. 9.1 and is further discussed in Section 9.3. The upward channel branching above the aircraft and the downward branching below the aircraft seen in Fig. 9.1 indicate that the discharge propagated away from the aircraft in both upward and downward directions, and hence that the lightning must have been initiated at and by the aircraft. A video of the event from which Fig. 9.1 is taken is found at www.crh.noaa.gov/pub/ltg/plane_japan.php.

Fig. 9.1   A video frame showing the initiation of cloud-to-ground lightning by an aircraft taking off from the Kamatsu Air Force Base in Japan during winter. Courtesy of Zen Kawasaki.

The history of lightning strikes to aircraft starts with lighter-than-air craft, so-called airships. The first recorded hot-air balloon flight took place in 1783, and is attributed to Joseph and Etienne Montgolfier, French paper manufacturers. The first hydrogen-filled balloon flight was apparently made by the French physicist Jacques Charles in the same year. Considerable information on lightning interaction with airships is available from the era of the large rigid (containing internal framework) dirigibles, which were used in the first third of the twentieth century for passenger transport and for military reconnaissance and bombing (see Section 9.3). The rigid airship's most important and best-known inventor was Count Ferdinand von Zeppelin whose dirigible-building motivation apparently originally came from observing lighter-than-air balloons reconnoitering Confederate activities during the US Civil War. President Abraham Lincoln had a small fleet of balloons built for this purpose. The 127 m (416 foot) long Luftschiff Zeppelin One (LZ 1) (Luftschiff is German for airship) had her maiden flight on July 2, 1900. The most famous airship, the Hindenburg, was designated LZ 129 (the 129th in the series), and, at 245 m (804 feet) in length, was the largest aircraft of any kind ever to fly. The Luftschiff Zeppelins built in Germany and the similar rigid airships manufactured in other countries were constructed of an extensive metal framework covered by a fabric envelope with the airship's lift being derived from hydrogen-filled compartments contained within the metal framework. The metal framework provided a Faraday cage to shield the hydrogen from contacting the hot lightning channel in the event of a lightning strike, and this shielding was generally successful. The three rigid hydrogen-filled airships that are thought to have burned because of lightning strikes were all apparently venting hydrogen at the time of the strikes, an activity to be avoided during flights in and around thunderstorms (Archbold 1994). In fact, the Hindenburg was struck by lightning a number of times with the only effect being holes up to 5 cm (2 inches) in diameter burned in the fabric covering the metal frame (Archbold 1994). The Hindenburg was destroyed in 1937 by a hydrogen fire while it was docking in Lakehurst, New Jersey, after a trans-Atlantic flight. Lightning was not involved, but the weather had been bad with heavy rain falling prior to the docking. It is generally thought that a corona discharge (St. Elmo's fire) on the top, rear airframe ignited the hydrogen from a leaking hydrogen-storage compartment in the vicinity of the corona. The airship burned from end to end in 34 s. Amazingly, 62 of the total 97 passengers and crew members survived the conflagration. Additional case histories of the interaction of lightning with lighter-than-air aircraft are found in Section 9.3.

There have been four major research programs involving airplanes that were intentionally flown into thunderstorms to be struck by lightning. The airplanes used in the programs were:

(1) An F-100F, a single-engine jet aircraft, as part of the US Air Force Cambridge Research Laboratories Rough Rider Project that took place from 1964 through 1966 (Fitzgerald 1967, Petterson and Wood 1968). The instrumented F-100F studied Florida thunderstorms. It penetrated the storms to measure turbulence and to obtain lightning photographic, shock wave, and electrical current records.

(2) An F-106B, a delta wing, single-engine jet aircraft of 21.5 m (about 70 feet) length including a sharp nose boom (a slender metal extension projecting from the plane's nose), as part of the NASA Storm Hazards Program that took place from 1980 through 1986 (e.g., Pitts *et al.* 1987, 1988). The F-106B flew through thunderstorms about 1500 times at altitudes ranging from 5000 to 40 000 feet (1.5 to 12 km) and was struck by lightning over 700 times. Almost ten times as many strikes were obtained for the high altitudes as for the low, although the number of cloud penetrations at high and low altitudes, a total near 1500 as noted above, was not much different. Statistics were compiled relating to the characteristics of the electric and magnetic fields measured on the aircraft surface and the lightning current flowing through the aircraft.

(3) A CV-580, a two-engine turboprop transport aircraft of 24.7 m (about 80 feet) length, as part of the US Air Force/Federal Aviation Administration Lightning Characterization Program that took place in 1984, 1985, and 1987 (e.g., Rustan 1986, Reazer *et al.* 1987, Lalande *et al.* 1999). The CV-580 was instrumented for detailed electric field, magnetic field, and current measurements.

(4) A C-160, a two-engine aircraft similar to the CV-580 but somewhat larger, as part of the French Transall Program that took place in 1984 and 1988 (e.g., Moreau *et al.* 1992, Lalande *et al.* 1999). The major part of the study was

conducted in 1988 in the south of France. The C-160 was instrumented specifically for study of the initial processes of lightning occurring at the aircraft, including a high-speed video camera.

A detailed discussion of all four airborne research programs is found in Uman and Rakov (2003).

The first direct evidence of the initiation of a lightning strike by an aircraft was provided by a ground-based research radar that observed radar echoes from lightning channels occurring during strikes to the NASA F-106B research aircraft (Mazur et al. 1984). The initial leader channels originated at or very near the F-106B (the radar resolution was 150 m) and extended away from it with increasing time. The much less common type of lightning strike, the interception of an already-existing lightning flash by the F-106B, was also observed by the ground-based radar (Mazur et al. 1986). The fact that aircraft generally initiate the lightning that strikes them was also inferred from the analysis of measured currents and electric field waveforms on the surfaces of the USAF/FAA CV-580 and the French C-160. Thirty-five of 39 strikes to the CV-580 were interpreted to be aircraft-initiated (Reazer et al. 1987). The C-160 obtained high-speed video records of channel formation that further supported the view that the aircraft initiated the lightning (e.g., Moreau et al. 1992).

In an ambient electric field of about $50 \, \text{kV} \, \text{m}^{-1}$, a not uncommon value in thunderstorms near the altitude of 16 000 feet (about 5 kilometers) at which the CV-580 and C-160 research aircraft flew, an aircraft similar to those two aircraft (according to the interpretation of the data acquired via the physical model proposed by Mazur [1989]) would initially launch a positively charged leader in the direction of the electric field from one extremity (wingtip, nose, tail) of the aircraft, where the ambient electric field was sufficiently enhanced by the small radius of the curvature of the conductor to allow the initiation. In Fig. 9.2, the positive leader is initiated from the nose of the aircraft at the time and field point A, after which the electric field on the aircraft increases as the aircraft charges negatively owing to the removal of positive charge by the leader. A few milliseconds after the positive leader is launched, when the aircraft electric field is raised sufficiently by the negative charging, a negatively charged leader is initiated (at point B) from the opposite extremity (the tail in Fig. 9.2) of the aircraft. It is reasonable to expect that a positive leader would occur first in the bidirectional leader development since, in general, positive leaders are initiated and can propagate in lower values of electric field than can negative leaders (as determined from laboratory experiments). After the positive leader initiation, the electric field at the aircraft would be increased both via the positive leader's removal of positive charge from the aircraft and via the elongation of the overall conducting system of the airplane and positive leader in the ambient field, thereby resulting in an electric field large enough for negative leader initiation from an aircraft extremity opposite to the extremity that initiated the positive leader. The aircraft extremities (wingtips, nose, and vertical stabilizer) provide the region of high electric field needed to initiate a lightning discharge by

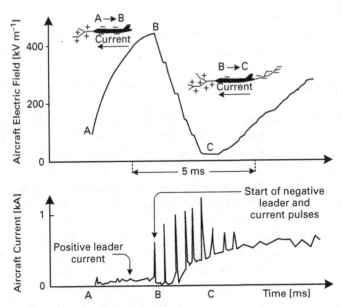

Fig. 9.2 An illustration of the mechanisms of lightning initiation by an aircraft in flight. Adapted from Lalande *et al.* (1999). Reprinted with permission from SAE paper 1999-01-2397 ©1999 SAE International.

enhancing the ambient electric field to the breakdown value, about $1.5 \times 10^6 \,\mathrm{V\,m^{-1}}$ at 5 km altitude. Thus, the aircraft enhancement factor (the factor that the ambient electric field is increased by being concentrated at the aircraft extremities) must be about 30 to initiate lightning in the observed ambient fields. After the negative leader initiation at point B in Fig. 9.2, impulsive currents with peaks near 1 000 amperes likely associated with the steps of the negative stepped leader are the dominant feature for a period of some milliseconds. Thereafter, after point C in Fig. 9.2, the observed current through the aircraft is generally composed of a steady component and a variety of impulses, probably not unlike a natural intracloud flash. Most aircraft-initiated lightning flashes are probably similar to natural intracloud lightning. Occasionally, aircraft initiate or otherwise become involved in cloud-to-ground lightning, this being more likely when they are closer to the Earth. Clearly, if an aircraft initiates lightning at low enough altitudes, such as soon after takeoff, as in the case shown in Fig. 9.1, that aircraft will necessarily be involved in a ground flash.

According to a United Air Lines study (Harrison 1965) of 99 lightning strikes, electrical discharges to airplanes in flight exhibit three common features: (1) a bright flash, sometimes blinding, (2) a loud explosive "boom," sometimes muffled, and (3) minor damage to the aircraft in one-third to one-half of the strikes.

Pilots often identify two different types of lightning–airplane interaction which they term "static discharge" and "lightning." The former, "static discharge," is characterized by radio static on the pilot's earphones of some seconds duration and a corresponding corona discharge (when it is dark, the corona is visible as luminous St. Elmo's fire) on the exterior aircraft surfaces before the major observed electrical discharge. The other type of lightning–airplane interaction, "lightning," is an electrical discharge that occurs without much prior warning. "Static discharges" are the much more common occurrence and apparently correspond to aircraft-initiated lightning. As noted above, they are usually similar to natural intracloud discharges. The "lightning" category apparently includes primarily flashes initiated independently of the aircraft with which the aircraft then interacts. Both "static discharge" and "lightning" do similar damage to the aircraft skin and interior electronics.

9.2   Statistics

Figure 9.3 summarizes the results of five studies of the altitude at which airplane/lightning incidents occur. These studies took place between the early 1950s and the mid-1970s. During the period 1950 to 1974, a typical US commercial aircraft was struck once every 3000 flight hours, or about once a year (Fisher *et al.* 1999). The statistics are similar for all types of aircraft. Older piston aircraft which cruise at 10 000 to 15 000 feet (about 3 to 4.5 km) show a strike pattern as a function of altitude similar to that of jet aircraft which cruise at much higher altitudes. For modern commercial jets, most strikes occur either in climbing to a cruising altitude, generally near 30 000 feet (about 9 km), or in landing, when the aircraft passes through the region of the cloud where the temperature is near the freezing level, 0 °C or 32 °F. Most strikes occur when the aircraft is within a cloud with only a few percent of strikes taking place when the aircraft is below or beside a cloud. The vast majority of strikes is associated with turbulence and precipitation. Although a typical thundercloud charge distribution is shown in Fig. 9.3, not all lightning strikes are associated with typical thunderclouds. For example, strikes have been recorded in clouds described as composed solely of ice crystals. The United Airlines report states that any weather situation producing precipitation appears to be capable of causing electrical discharges to aircraft in flight.

Statistics on over 1000 lightning strikes to commercial jets in Japan are shown separately for summer and winter in Fig. 9.4a and b, respectively (Murooka 1992). Most of the strikes in summer and winter occur in the same temperature range, −5 to 0 °C, although that temperature range is much lower in altitude in the winter (in fact, it is very near the ground). The strike data from summer and winter are combined in Fig. 9.5 where they are plotted against ambient temperature.

South African Airways lightning strike records indicate that most lightning incidents occurred 3 to 5 km (about 9800 to 16 400 feet) above sea level (Anderson and Kroninger 1975). The number of strikes reported per 10 000 hours of flying time for different years (1948 to 1974) varied between about 1 and 4, consistent with the US data discussed above.

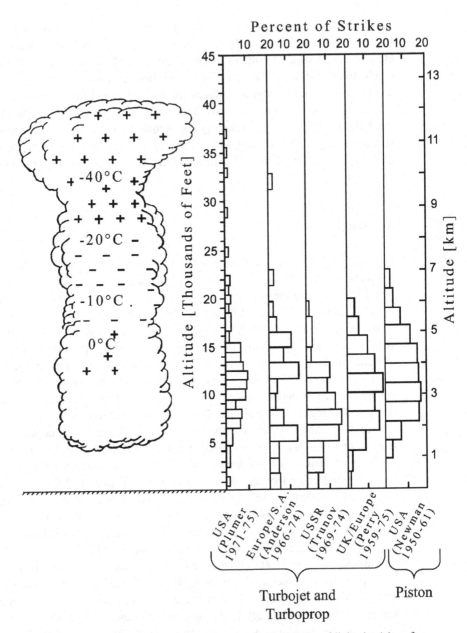

Fig. 9.3   Histograms from five different studies showing the incidence of lightning/aircraft interactions as a function of altitude: USA 1971–1975; Europe/South Africa 1966–1974; USSR 1969–1974; UK/Europe 1959–1975; USA 1950–1961. A typical thunderstorm charge distribution is shown at left. Adapted from Fisher *et al.* (1999) and Rakov and Uman (2003).

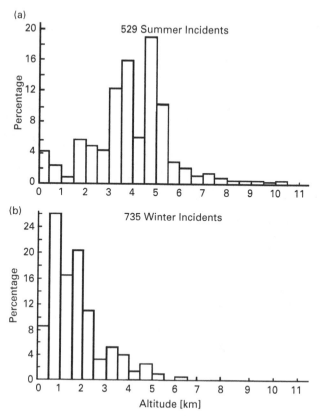

Fig. 9.4   (a) Lightning/aircraft incidents vs. altitude for commercial aircraft in summer in Japan, (b) in winter in Japan. Adapted from Murooka (1992).

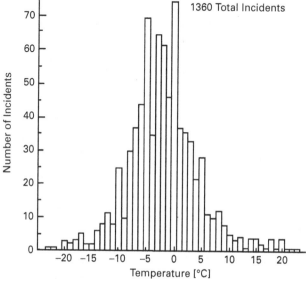

Fig. 9.5   Lightning/commercial aircraft incidents vs. ambient temperature at the altitude of the incident for all seasons in Japan. Adapted from Murooka (1992).

Fig. 9.6   Radome damage caused by a lightning strike to an aircraft near Los Angeles on January 28, 1969. Photo courtesy of the Los Angeles Times.

The effects of lightning on aircraft are generally minimal, although the consequences of the interaction can occasionally be catastrophic, as we shall see in Section 9.3. Lightning damage is usually divided into "direct" and "indirect" (indirect is often called "induced") effects. Direct effects occur at the points of the lightning contact and include the puncturing or splintering of non-metallic structures such as the plastic radomes that cover the radars located at the front of aircraft (Fig. 9.6), the burning of holes in metal skins (Figs. 9.7–9.9), the welding or roughening of movable hinges and bearings, damage to antennas and lights located at aircraft extremities, and fuel ignition. The radome shown in Fig. 9.6 is an electrically insulating (as opposed to conducting) cover that protects the aircraft's radar from wind and weather but allows the radar signal to pass through it unimpeded. The radome has thin metallic "diverter" strips across it that are intended to shunt the lightning current to the metal fuselage but not to interfere with the transmission or reception of the radar signal. The diverter strips apparently did not perform as designed. Indirect effects are those produced by the deleterious voltages and currents induced within the aircraft by the lightning electric and magnetic fields that enter through openings in the aircraft's metal skin or are induced on antennas. Indirect effects include upset or damage to any of the many aircraft electronic systems. According to Fisher *et al.* (1999), 20 percent

Table 9.1 Incidence of indirect effects in commercial aircraft during 214 lightning strikes.

	Interference	Outage
HF communication set	–	5
VHF communication set	27	3
VOR receiver	5	2
Compass (all types)	22	9
Marker beacon	–	2
Weather radar	3	2
Instrument landing system	6	–
Automatic direction finder	6	7
Radar altimeter	6	–
Fuel flow gauge	2	–
Fuel quantity gauge	–	1
Engine rpm gauges	–	4
Engine exhaust gas temperature	–	2
Static air temperature gauge	1	–
Windshield heater	–	2
Flight director computer	1	–
Navigation light	–	1
AC generator tripoff	6	
Autopilot	1	–

Adapted from Fisher *et al.* (1999).

of 851 reported aircraft strikes in the United States resulted in indirect effects. Examples of the instruments affected are given in Table 9.1. In the South African study referred to above, aircraft frame or instrument damage occurred in 40 percent of the 245 recorded aircraft strikes.

9.3 Accidents

Kapitänleutnant Martin Dietrich was returning to Germany from a World War I bombing raid on England in Luftschiff Zeppelin 42 (LZ 42, also called *L 42*) when the airship encountered lightning, as described by Robinson (1971).

On the homeward flight *L 42* found a wall of black thunderclouds reaching up to 23,000 feet barring her way. Dietrich had no choice but to fly through them. With antenna wound in, and pressure height of 19,400 feet, *L 42* plunged into the black storm clouds at 16,400 feet. Hail drummed on the taut outer cover, and at 4.45 a.m. a blinding flash of lightning struck the ship. The metal structure was so heavily charged that a machinist's mate, sitting on a stool in the port midships engine car, got a severe shock when he touched the duralumin gondola wall. There was a strong smell of ozone in the rear gondola, and its personnel believed that the electrical charge left the ship along the port propeller bracket. Ten minutes later the Zeppelin was staggered by a second lightning bolt. This time the top lookout saw it strike near him and course along the ship's back, while people in the rear gondola saw the flash shooting out of the tail. The sailmaker patrolling the keel was astonished to see the lightning glaring through the translucent gas cells and outer cover. Fifteen minutes later there was a weaker lightning stroke which was seen from the control car to hit forward. A meticulous check of the ship at Nordholz revealed six holes in the cotton outer cover at the bow, the largest the size of the palm of a hand. Underneath, two bracing wires in contact had burned through, and a pea-sized hole had been punched in a duralumin girder member. Traces of fire were found on the port after propeller.

Van Orman (1978) gives a first-person account of three hot-air balloons being struck by lightning near Pittsburgh, PA, during the National Balloon Race of 1928. The lightning strike to the balloon he was piloting contacted the balloon near its equator and followed the envelope from that point down to the crew basket, passing through the basket and killing his aide. The balloon exploded leaving only a piece of the top fabric to act as a parachute. Van Orman landed with only a broken ankle but with a dead companion. Interestingly, the balloon had virtually no metal components except for a few hinges and screws. The envelope was rubberized cotton fabric contained within a cotton seine net with the wicker crew basket suspended beneath by natural fiber manila ropes. Van Orman (1978) states that it is his opinion that lightning will not strike a dry balloon, but that wet ones are vulnerable, presumably because they are electrically conducting to some extent. That seems reasonable, but certainly not proved.

Lightning can cause damage to modern heavier-than-air aircraft that varies from minor pitting of the aluminum skin to complete destruction of the aircraft. Apparently, weather conditions can sometimes be conducive to making lightning triggering by aircraft more likely and then even multiple aircraft may be involved. This was the case in January 28, 1969, when four separate aircraft were struck by lightning near Los Angeles (Los Angeles Times 1969), one of which is shown in Fig. 9.6. This was also the case on February 24, 1987, when at least six aircraft were struck by lightning while arriving or departing airports in the Los Angeles area in a period of a few hours. The winter storm system present that day produced rain showers and occasional lightning. Four Boeing 727s, flying between 3800 and 8000 feet (between about 1.1 and 2.4 km), had lightning-caused holes punched in their radomes, and a Boeing 737 suffered unspecified damage at 3200 feet (about 1 km). A NASA T-38A jet flown by two astronauts suffered a lightning-induced in-flight explosion at 2500 feet (about 0.75 km) followed by a fire that extensively damaged the center fuselage. The T-38A, still on fire, landed at a military base near Los Angeles. Fortunately, the crew escaped injury. The official report describing the T-38A incident is found in McMurtry (1987). Another similar group lightning strike, although much less well documented, occurred on February 19, 2006, when four separate aircraft, all operated by RyanAir, were struck descending over the northern coast of Spain. All planes landed safely.

On December 8, 1963, a Pan American World Airways Boeing 707 was flying in a holding pattern at 5000 feet (about 1.5 km) near Elkton, Maryland. Ninety-nine witnesses reported a cloud-to-ground lightning flash near or on the aircraft at about the time it was seen bursting into flames. All 73 passengers and eight crew members were killed. An investigation determined that three fuel tanks had exploded and that

Fig. 9.7 Photograph of the left wingtip of the Boeing 707 that was destroyed by lightning on December 8, 1963, near Elkton, Maryland. On the model insert, the white portion of the wing gives the relative size and orientation of the portion of the actual wing shown. A number of lightning-caused holes and considerable pitting are evident. Courtesy of Bernard Vonnegut and Roger Cheng.

Fig. 9.8   A closer view of the Boeing 707 wingtip shown in Figure 9.7. Five lightning-caused holes are visible. Courtesy of Bernard Vonnegut and Roger Cheng.

there were lightning strike marks and holes on the left wingtip. Photographs of this lightning damage are reproduced in Figs. 9.7–9.9. Evidence indicated that the left reserve fuel tank, the outermost fuel tank in the left wing, exploded first, followed by the center and right reserve fuel tanks. There was lightning damage about 30 cm (about one foot) from the edge of the vent outlet of the left reserve fuel tank. The largest single indication of lightning was an irregular-shaped hole about 4 centimeters in diameter burned through the top of the wing, shown best in Fig. 9.9. After the accident investigation and as a result of additional research, the required thickness of the aluminum skin enclosing the fuel on 707s and on other aircraft was increased to reduce the likelihood of burn-through, and fuel filler caps (similar to the gas cap on your car) and fuel access plates were required to be better bonded to the airframe in order to inhibit potential sparking in the vicinity of fuel. The official report (Aircraft Accident Report, Boeing 707–121 N709PA Pan American World Airways, Inc., near Elkton, Maryland, December 8, 1963, Civil Aeronautics Board File No. 1-0015, February 25, 1965) attributes the disaster to "lightning-induced ignition of the fuel/air mixture in the No. 1 reserve fuel tank with resultant explosive disintegration of the left outer wing and loss of control."

On May 9, 1976, an Imperial Iranian Air Force B-747, Flight ULF48, was struck by lightning near Madrid, Spain, with a catastrophic result. The aircraft was on a flight to the United States from Iran, with an intermediate stop in Madrid. The last radio contact was made as the aircraft was descending to 5000 feet (about 1.5 km) in

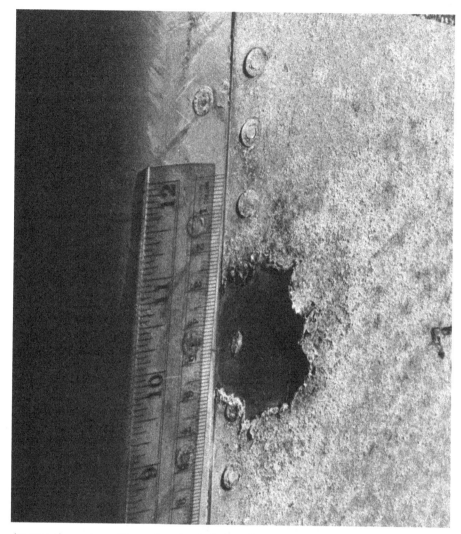

Fig. 9.9   An even closer view of the major wingtip hole and surrounding pitting seen in Fig. 9.7 and Fig. 9.8. Courtesy of Bernard Vonnegut and Roger Cheng.

clouds, probably near an altitude of 6000 feet (about 1.8 km). Since the aircraft involved, a Boeing 747, was used extensively in commercial operations worldwide (and still is), and, in view of the nature of the accident, the US National Transportation Safety Board requested and was granted permission to assist in the investigation. The resultant report is labeled NTSB-AAR-78-12, October 1978: Special Investigation Report-Wing Failure of Boeing 747-131, Near Madrid, Spain, May 9, 1976, from which the discussion below is taken.

Two witnesses reported seeing lightning strike the aircraft. Some witnesses said they saw an in-flight fire confined to the No. 1 engine. Other witnesses reported seeing an in-flight explosion and fire followed by the separation of aircraft parts. Pitting and localized burn areas typical of lightning attachment damage were found on the left wingtip and on the vertical fin. No holes were burned through the aircraft skin into any of the fuel tanks. The left wing had separated into 15 major pieces before ground impact and parts of it were found at a number of locations.

Several motor-operated valves were present in the fuel tanks, and the electric motors that operated these valves were mounted on the outside surfaces of the front or rear wing spar. The motors were connected to the valves by mechanical couplings or drive shafts which penetrated the spars. The motor for the valve in the No. 1 fuel tank was never recovered. The drive shaft was found and was determined to be electrically insulated at the spar penetration. The mechanical coupling/drive-shaft arrangement may have provided a path for an electric current to enter the tank and cause a spark in a fuel/air mixture. The level of residual magnetization in this area of the valve was indicative of high currents in that part of the plane.

The evidence (1) that the explosion in the No. 1 tank occurred in the immediate area of a motor-driven fuel valve, (2) that the motor was never recovered, (3) that a high level of residual magnetization existed in the ferrous material in this area, (4) that certification tests showed this area to be a likely lightning-attachment point, (5) that lightning strikes are known to have disabled the motors on other aircraft, and (6) that no other possible ignition source could be determined, provided the foundation for the hypothesis that the tank explosion was ignited by a spark at this motor-driven valve.

The official report (NTSB-AAR-78-12) states that

assuming that a lightning strike can generate a source of ignition to fuel vapors, aircraft fuel explosions could occur more frequently. However, events must combine simultaneously to create the explosion, and this combination would occur rarely. In this case, the events were (1) an intermittently conductive path which closed and opened an electrical loop, (2) a lightning-induced current of sufficient intensity flowed in this path and formed a spark, and (3) a flammable vapor surrounded this spark. Possibly this combination of events has occurred a number of times before, in the following accidents: (a) Milan, Italy (Constellation); (b) Elkton, Maryland (B-707); (c) Madrid, Spain (USAF KC-135); (d) KSC, Florida (USAF F-4); (e) Pacallpa, Peru (L-188).

Accident (b) is discussed above.

In November 2005 the Federal Aviation Administration (FAA) issued a Notice of Proposed Rulemaking that would establish a set of requirements that do not specifically require the "inerting" of the fuel in aircraft fuel tanks (that is, rendering the fuel incapable of exploding) but rather set acceptable levels of flammability exposure in fuel tanks. The FAA stated that 17 aircraft had been destroyed since 1960 as a result of fuel tank explosions (including four aircraft since 1989). These explosions were due to electrical discharges in the fuel tank, some of which were lightning. Probably the most spectacular recent such event involved TWA 800, a Boeing 747, which exploded off Long Island, New York in 1996 soon after takeoff. The explosion was attributed to a wiring short-circuit producing a spark in the center fuel tank that was purposely left nearly empty for the flight. Apparently, the nearly empty fuel tank contained a fuel–air mixture that was flammable. Experimental inerting systems for fuel tanks have been recently flown on Boeing 747s and 737s and on Airbus A320s. Among the most straightforward inerting systems are those that replace the air (which is about 20 percent oxygen) in the fuel tank with nitrogen since no fire or explosion can occur in the absence of oxygen.

On October 15, 1965, a Convair aircraft, Flight 517, was in the process of taking off from the Salt Lake City Airport. There was some light rain in the area but apparently there was no lightning other than the event to be described. During takeoff, an extremely loud noise was heard. The first officer also observed a blue-white glow around the nose of the aircraft at the time of the noise. Observers in the control tower confirmed that there was a lightning strike to the plane. The aircraft returned to the airport. Three large holes were found in the runway which matched the exact dimensions of the two main landing gear and the nose wheel. The largest hole, under the right main gear, was nearly 2 m in diameter and 15 to 20 cm deep. Pieces of asphalt as large as 0.3 m had been hurled 30 to 50 m down the runway. The aircraft suffered numerous burns to the wheel rims and fuselage just behind the nose wheel-well. The rotating beacon, the grounding wire on the right main gear, and the fixed vertical stabilizer cap were burned off. The fact that there was little if any lightning in the area at the time of the strike to the Convair implies that the aircraft initiated the lightning.

On February 8, 1988, a Fairchild Metro III commuter airliner powered by two turboprop engines and carrying 19 passengers and two crew members on a flight from Hannover to Düsseldorf, Germany, was struck by lightning and subsequently crashed, killing all on board. The Fairchild Metro III was approaching Düsseldorf at an altitude of about 3000 feet (about 0.9 km). The pilot had just lowered the landing gear when the plane fell and rose in altitude between 2500 and 3000 feet (between about 0.75 and 0.9 km) as the pilots tried to trim the aircraft for proper descent. As they were stabilizing the aircraft, lightning struck it and apparently disconnected all batteries and generators from the aircraft's electrical system, also terminating the cockpit voice recorder record. Without electrical power, the pilots evidently had no control of the landing gear and limited control of the flaps. The aircraft was inside a cloud and had no cockpit lights so the pilots would probably not have been able to read their instruments. Emergency flashlights apparently were not present in the aircraft as required, or at least none was found at the crash scene. Observers on the ground saw the aircraft dive out of the cloud base and then climb again into the cloud, this pattern being repeated two or three times. On one of these oscillations in altitude, the right landing gear was torn from the aircraft, further destabilizing it. The subsequent aircraft motion resulted in a wing being separated from the aircraft. The Fairchild went into a spiral dive and crashed. A reconstruction of the electrical system failure pointed to the failure of a critical relay. The official report of the accident is found in "Bericht über die Untersuchung des Flugunfalles mit dem Flugseug SA Z27-AC, Metro III, D-CABB, am 8. Februar 1988 bei Kettwig AZ.: 1X001/88, Flugunfallenuntersuchungsstelle beim Luftfahrt-Bundesamt, Bundesrepublik Deutschland."

On February 26, 1998, a US Airways Fokker F28 MK 0100 flying from Charlotte, NC, to Birmingham, AL, carrying 87 passengers and five crew members was struck by lightning with no immediate effect. However, within a few minutes the aircraft suffered a failure of both its hydraulic systems. In order to make an emergency landing, the landing gear and flaps were extended via an alternate method but without control of the nose landing gear steering. A number of brake applications were also possible in the absence of the hydraulic systems. On landing, the aircraft traveled about 330 m in the grass off the left side of the runway. The nose landing gear separated from the aircraft and the nose section came to rest on a taxiway about 160 m from the aircraft. Airport personnel reported finding pieces of the main landing gear tires on the runway, and the shimmy damper reservoir for the left main landing gear was found on the left side of the runway. Examination of the two hydraulic system reservoirs of the airplane revealed both were empty and hydraulic fluid was noted on the vertical stabilizer. When the hydraulic systems were pressurized, leakage occurred from a hole in the No. 1 elevator pressure line approximately three-quarters of the way up the vertical stabilizer and from a second hole in the No. 2 elevator return line, this hole being located behind the rudder flutter damper approximately half way up the vertical stabilizer. Examination of the airframe revealed that the right exterior fuselage skin exhibited approximately 103 lightning burn marks which ranged in size from 0.16 to 1.6 cm (1/16 inch to 5/8 inch) in diameter. Additionally, the right stabilizer showed evidence of scorching at the outboard corner of the upper surface at the trailing edge. The outboard static wick on the right stabilizer was missing with evidence of heat at its base. Additionally, a bonding strap that provided an electrical connection between the horizontal and vertical stabilizers failed and the strap was discolored. Apparently, lightning current flowing in the bonding strap between the vertical and horizontal stabilizers side-flashed to the hydraulic lines, burning through them and releasing the hydraulic fluid. A report on this accident by the US National Transportation Safety Board is found at www.ntsb.gov/aviation/MIA/98A089.htm.

In addition to the variety of lightning interactions with airships and airplanes, examples being given above, there have been high-profile lightning strikes to the Apollo 12 and Atlas-Centaur 67 launch vehicles soon after they left their launch pads.

On November 14, 1969, the Apollo 12 space vehicle was launched from the NASA Kennedy Space Center (KSC) on the first phase of its trip to the Moon. Major electrical disturbances occurred within a minute of liftoff that were later determined to be due to two separate vehicle-initiated lightning strikes, one resulting in a cloud-to-ground lightning, the other in an intracloud lightning. Nine non-essential instrumentation sensors were permanently damaged. There was momentary loss of communications, instrument readings were disturbed, various warning lights illuminated and alarms sounded in the crew compartment, three fuel cells disconnected from their buses, the inertial platform lost altitude reference, and various clocks malfunctioned. All critical system problems were corrected when the spacecraft reached Earth orbit, and the mission successfully delivered two astronauts to the surface of the Moon and returned them to Earth. At the time of launch

(11:22 a.m. EST), a cold front was passing through the launch area. Isolated thunderclouds within 50 km of KSC reached a maximum height of 23 000 feet (about 7 km). In the vicinity of the launch complex, broken clouds were reported at 800 feet (about 0.25 km) with a solid overcast from about 10 000 to 21 000 feet (about 3 to 6 km). The freezing level was near 12 400 feet (about 3.8 km). No lightning was reported in the KSC area six hours prior to or after the launch, although the instrumentation available for detecting lightning was primitive.

The vehicle apparently initiated a cloud-to-ground lightning discharge 36.5 seconds after launch at an altitude of about 6400 feet (about 1.9 km) and then triggered an intracloud discharge at 52 seconds at an altitude of about 14 400 feet (about 4.4 km). In the 20 minutes prior to launch the vertical electric field at ground near the launch site was rapidly varying, but the crude electric field measuring devices (radioactive probes, later to be replaced by field mills – see Section 8.1) used at the time were not calibrated, so the actual field magnitudes were not known. The possibility that the Apollo vehicle could initiate lightning had not been previously considered, according to Godfrey *et al.* (1970), the official report that presents the findings of the team that investigated the incident. The realization that Apollo 12 had initiated lightning led to a very significant round of funding for research into triggered and natural lightning and for the development of a variety of modern instruments to monitor the electrical characteristics of clouds and to determine lightning locations and characteristics. According to the calculations found in Godfrey *et al.* (1970), if a Saturn V vehicle 300 m long (including the total exhaust plume) with a 5 m radius and a 10 cm radius-of-curvature top cap were placed in an electric field of $7.5 \, \text{kV} \, \text{m}^{-1}$, the field at the top cap would be enhanced 320 times to produce a breakdown field of $2.4 \times 10^6 \, \text{V} \, \text{m}^{-1}$ at an altitude of 6000 feet (about 1.8 km). The Saturn vehicle was 110 m long, its opaque exhaust was about 40 m, and its total visible exhaust was about 200 m (Krider *et al.* 1974). Commonly observed field values in thunderstorms are 50 to $100 \, \text{kV} \, \text{m}^{-1}$, but the clouds present might not have had fields of such a high value. Once the field at the pointed upper extremity of the vehicle exceeded the breakdown field, a positive discharge would have emanated from that location of the vehicle toward the cloud charge, assuming the overhead cloud charge was negative. This positive leader then further enhanced the electric field both at the tip of the upward-propagating discharge and at the opposite end (or exhaust) of the vehicle, resulting in a downward-propagating negative-stepped leader from the exhaust. For the case of the first lightning initiated by Apollo 12, and the one event initiated by Atlas-Centaur 67 (to be discussed next) these two vehicle-initiated discharges were apparently not dissimilar in their characteristics from natural downward cloud-to-ground lightning flashes.

On March 26, 1987, the Atlas-Centaur 67 vehicle was launched from the Cape Canaveral Air Force Station, Florida, adjacent to the Kennedy Space Center. Weather conditions during the late afternoon were similar to those at the time of the Apollo 12 launch. There was a broad cloud mass covering most of Florida and the Gulf of Mexico, and a nearly stationary cold front, oriented southwest–northeast, extending across northern Florida well north of Cape Canaveral. A weak squall line, also oriented southwest–northeast, was centered over the eastern Gulf of Mexico and was moving eastward over the Florida peninsula. This squall line produced substantial amounts of cloud-to-ground lightning activity of both negative and positive polarity throughout the day, but almost without exception this activity was well west of the Cape. At the launch site there was heavy rain, and layer clouds were reported at altitudes between 8000 and 20 000 feet (about 2.4 and 6.1 km). No cloud-to-ground lightning had been observed within 5 nautical miles (9.3 km) of the launch site in the 42 minutes prior to launch and only one discharge was within 10 nautical miles (18.5 km) during this time. A cloud discharge apparently occurred about 2 minutes prior to launch, undetected by the KSC lightning detection instrumentation, but reported to the author after the launch by members of the press corps. At the time of launch, the electric field at the launch site was negative $7.8 \, kV \, m^{-1}$, a value indicating substantial negative charge overhead. There were no constraints on the unmanned launch relative to the electric field value allowed for launch. Such constraints had been developed for manned launches after Apollo 12 limiting allowable launch fields to about 25 percent of the field present at the time of Atlas-Centaur 67 launch. Forty-nine seconds after launch, when the vehicle was at an altitude of about 12 000 feet (about 3.6 km), a lightning flash was observed below cloud base. That flash produced at least four strokes to ground which were recorded by television cameras. The first two strokes that could be resolved on the video records followed one channel to ground, while the latter two followed two separate channels to ground yielding a total of three ground strike-points. At the time of lightning initiation, the vehicle was at a height near 12 000 feet (about 3.6 km) where the temperature was +4 °C, while the freezing level was at 14 400 feet (about 4.4 km). The precipitation inside the cloud, as measured by radar, was far less than characteristic of thunderstorms in that area, the precipitation level generally being related to cloud charge generation and separation. From the magnetic field signal recorded by the KSC lightning locating system, the first stroke current was determined to be of negative polarity and was estimated to have a peak value of 20 kA, a common value for first strokes in natural lightning.

At the time of the lightning strike there was a memory upset in the part of the vehicle guidance system called the digital computation unit, leading to an unplanned vehicle rotation. The stresses associated with this motion caused the vehicle to begin breaking apart. About 70 seconds after liftoff, the range safety officer ordered the Atlas-Centaur destroyed. Substantial portions of the fiberglass-honeycomb structure that covered the front 6 to 7 m of the vehicle were subsequently recovered from the Atlantic Ocean. These showed physical evidence of lightning attachment. Approximately 40 percent of the telemetry outputs from the vehicle to the control headquarters exhibited anomalous electrical behavior at the time of the event. The Atlas-Centaur vehicle, which was about 40 m in length, served to enhance any electrical field in which it was immersed by about a factor of 30 to 50 (Bussey 1987). Thus a breakdown field of near $2 \times 10^6 \, V \, m^{-1}$ would exist at the nose of the vehicle in an ambient field of 50 to 80 $kV \, m^{-1}$, reasonable values in a thundercloud.

All of the information given in this section and further details of the Atlas-Centaur 67 event including reference to lightning involvement with two earlier Atlas-Centaur vehicles are found in Bussey (1987), the official report on the incident, and in Christian *et al.* (1989). As noted in Section 2.1, the dollar loss from the rocket and payload was estimated at $191 million.

9.4 Lightning test standards

Five standards for the lightning protection of aircraft have been written by the Society of Automotive Engineers (SAE) in coordination with the European Organization for Civil Aviation Equipment (EUROCAE) and are listed below:

1. ARP5412 Rev A published 2005 (ED 84), *Aircraft Lightning Environment and Related Test Waveforms*, www.sae.org
2. ARP5413 published 1999 (ED 81), *Certification of Aircraft Electrical/Electronic Systems for the Indirect Effects of Lightning*, www.sae.org
3. ARP5414 Rev A published 2005 (ED 91), *Aircraft Lightning Zoning*, www.sae.org
4. ARP5577 published 2002 (ED-113), *Aircraft Lightning Direct Effects Certification*, www.sae.org
5. ARP5416 published 2005 (ED-105), *Aircraft Lightning Test Methods*, www.sae.org

The acronym ARP stands for Aerospace Recommended Practices and the acronym ED stands for EUROCAE Document. Some of the ARP and the equivalent ED documents listed above differ in the material in their Appendices. There are numerous earlier versions of these standards published by the SAE, FAA, EUROCAE, and various military and other organizations, many of which are referenced in the documents listed above.

The first document, ARP5412, specifies a series of idealized voltage and current waveforms with which aircraft are to be tested for the effects of lightning. The idealized test current waveforms are labeled A, B, C, D, D/2, and H and are illustrated in Fig. 9.10. From the known characteristics of lightning, components A through D/2 represent severe currents in the strokes (first and subsequent) of cloud-to-ground lightning. ARP5412 also specifies the allowed approximations to the idealized waveforms of Fig. 9.10 that can be used in the laboratory for the actual testing. Component H is derived from the airborne F-106B, CV-580, and C-160 measurements discussed in Section 9.1 and is intended to describe the multiple current bursts observed when an aircraft in flight initiates lightning. The H-component can probably best be viewed as a conservative test waveform that accounts for both the current pulse bursts associated with negative stepped leaders from the initiation processes at the aircraft and the current pulses flowing through the aircraft from so-called recoil streamers originating at a distance from the aircraft when the leader from the aircraft propagates into highly charged regions of the cloud.

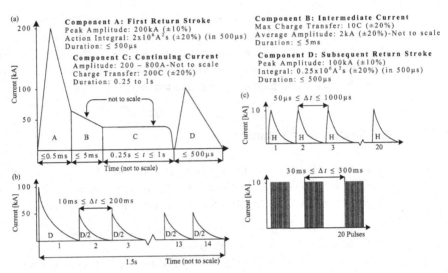

Fig. 9.10   Laboratory current waveforms specified for aircraft testing. (a) Current components A through D simulating the first two strokes in a cloud-to-ground flash, (b) multiple-stroke waveform simulating the second and additional strokes, and (c) multiple-burst (H-component) waveform showing the pulse waveshape (top) and pulse burst structure (bottom). Reprinted with permission from SAE ARP5412 Revision A © 2005 SAE International.

Four test voltage waveforms (not illustrated here) are found in ARP5412 that are intended to identify lightning attachment points and dielectric breakdown paths through non-conducting surfaces or structures. The voltages are to be imposed between an external electrode and the grounded airframe. The four different waveforms have been chosen for their ability to produce various forms of damage previously observed on aircraft involved with lightning. The voltage in ARP5412 waveform A increases linearly until breakdown occurs, waveform B has a 1.2 μs rise time to peak and a 50 μs time to half peak value, waveform C is a linear rising voltage chopped to zero at 2 μs, and waveform D has a rise time to peak of 50 to 250 μs and a time to half value of about 2 ms.

Different zones of the aircraft are expected to experience different levels of lightning severity, and these zones are defined in ARP5414. Zone 1A is that portion of the aircraft that can be expected to encounter a direct first return stroke but not the remainder of the flash because the forward speed of the aircraft results in a backward motion of the channel across the aircraft surface. Zone 1B is expected to encounter a first stroke and the remainder of the flash. Zone 1C is expected to encounter a reduced amplitude first stroke only and not the remainder of the flash. A subsequent stroke is likely to be swept into zone 2A by aircraft motion but with low probability of the remainder of the flash occurring there, while zone 2B is likely to encounter a swept subsequent stroke and the remainder of the flash. Zone 3 is expected to receive current conducted through the airframe, not direct lightning channel attachment. Different test current and voltage waveform components are to be applied to the different zones, as specified in Table 3 of ARP5412, for testing against both direct and induced effects. Procedures to test for indirect effects are considered in ARP5413.

References

Archbold, R. 1994. *Hindenburg: An Illustrated History.* New York: Warner Books.

Anderson, R. B. and Kroninger, H. 1975. Lightning phenomena in the aerospace environment. Part II: Lightning strikes to aircraft. *Trans. S. Afr. Inst. Electr. Eng.* **66**: 166–175.

Bussey, J. 1987. *Report of Atlas/Centaur-67/FLTSATCOM F-6 Investigation Board,* Vol. II. NASA.

Christian, H. J., Mazur, V., Fisher, B. D. *et al.* 1989. The Atlas/Centaur lightning strike incident. *J. Geophys. Res.* **94**: 13169–13177.

Fisher, F. A., Plumer, J. A., and Perela, R. A. 1999. *Lightning Protection of Aircraft,* second printing. Pittsfield, MA: Lightning Technologies Inc.

Fitzgerald, D. R. 1967. Probable aircraft "triggering" of lightning in certain thunderstorms. *Mon. Weather Rev.* **95**: 835–842.

Godfrey, R., Mathews, E. R. and McDivitt, J. A. 1970. *Analysis of Apollo 12 Lightning Incident. NASA* MSC-01540.

Harrison, H. T. 1965. United Air Line turbojet experience with electrical discharges. *United Air Lines Meteorological Circular,* No. 57.

Krider, E. P., Noggle, R. C., Uman, M. A. and Orville, R. E., 1974. Lightning and the Apollo/17 Saturn V exhaust plume. *J. Spacecraft Rockets* **11**: 72–75.

Lalande, P., Bondiou-Clergerie, A. and Laroche, P. 1999. Studying aircraft lightning strokes. *Aerospace Engineering* (publisher: SAE Aerospace): 39–42. *See also* Analysis of available in-flight measurements of lightning strikes to aircraft, *Proc. 1999 Int. Conf. Lightning and Static Electricity, Toulouse, France,* pp. 401–408.

Los Angeles Times, January 29, 1969. final edition, page 1.

Mazur, V. 1989. A physical model of lightning initiation on aircraft in thunderstorms. *J. Geophys. Res.* **94**: 3326–3340.

Mazur, V., Fisher, B. D. and Gerlach, J. C. 1984. Lightning strikes to an airplane in a thunderstorm. *J. Aircraft* **21**: 607–611.

Mazur, V., Fisher, B. D. and Gerlach, J. C. 1986. Lightning strikes to a NASA airplane penetrating thunderstorms at low altitudes. *J. Aircraft* **23**: 499–505.

McMurtry, T. C. 1987. *NASA 914 T-38A Jet Trainer Lightning Strike Investigation Report. Date of Mishap Feb. 24, 1987.* NASA Johnson Space Center document, July 6, 1987.

Moreau, J.-P., Alliot, J.-C., and Mazur, V. 1992. Aircraft lightning initiation and interception from in situ electric measurements and fast video observations. *J. Geophys. Res.* **97**: 15903–15912.

Murooka, Y. 1992. A survey of lightning interaction with aircraft in Japan. *Res. Lett. Atmos. Electr.* **12**: 101–106.

Petterson, B. J. and Wood, W. R. 1968. *Measurements of Lightning Strokes to Aircraft.* Sandia Laboratory Report SC-M-67–549. Albuquerque, NM: Sandia Laboratories. (also Report DS-68-1 of the Department of Transportation, Federal Aviation Administration, Washington, DC).

Pitts, F. L., Perala, R. A., Rudolph, T. H., and Lee, L. D. 1987. New results for quantification of lightning/aircraft electrodynamics. *Electromagnetics* **7**: 451–485.

Pitts, F. L., Fisher, B. D., Vladislav, V. and Perala, R. A. 1988. Aircraft jolts from lightning bolts. *IEEE Spectrum* **25** (July): 34–38.

Rakov, V. A. and Uman, M. A. 2003. *Lightning, Physics and Effects*. Cambridge University Press.

Reazer, J. S., Serrano, A. V., Walko, L. C. and Burket, Capt. H. D. 1987. Analysis of correlated electromagnetic fields and current pulses during airborne lightning attachments. *Electromagnetics* **7**: 509–539.

Robinson, D. H. 1971. *The Zeppelin in Combat*, 3rd edn. J. W. Caler Publishing Company.

Rustan, P. L. 1986. The lightning threat to aerospace vehicles. *AIAA J. Aircraft* **23**: 62–67.

Uman, M. A. and Rakov, V. A. 2003. The interaction of lightning with airborne vehicles. *Prog. Aerospace Sci.* **39**: 62–81.

Van Orman, W. T., and as told to Hull, R. 1978. *The Wizard of the Winds*. Saint Cloud, MN: North Star Press.

10 Ships and boats

10.1 History

Ships, the name given to large sea-worthy vessels, are generally made of electrically conducting metal, whereas boats, smaller vessels including speedboats, power-boats, motorboats, rowboats, and sailboats, are more often constructed of electrically insulating material such as fiberglass or wood. The metal shells of modern ships can be considered to be both an approximate Faraday cage and the outer surface of a topological shielded system (see Section 3.1), with the contact region between the electrically conducting hull and the ocean providing the grounding connection, and the ocean being the "ground." Thus, modern ships do not suffer much lightning damage. What damage does occur is generally limited to exposed communication antennas, radars, and insulating covers for equipment. On the other hand, wood and fiberglass boats seldom encounter lightning without being damaged in some way. The US lightning protection standard NFPA 780:2004 devotes Chapter 8, Protection for Watercraft, to the methods of protection for powerboats and sailboats.

Ten years or so after Benjamin Franklin proposed a method for the lightning protection of both houses and ships (see Section 4.1), those principles were applied to the protection of wooden ships, wood being the only material from which ships were constructed at that time. The history of the lightning protection of the ships of the British Royal Navy is particularly interesting and has been reviewed by Bernstein and Reynolds (1978), from which some of the following discussion is taken. In 1762, William Watson, one of Britain's early electrical scientists, wrote to the First Lord of the Admiralty recommending the installation of a lightning protection system on British Royal Navy vessels. Watson (1761) had previously suggested protecting sailing ships by connecting a brass wire conductor about "the thickness of a large goose quill" to the masts, and leading it from there, by the most convenient path, into the water. The British Royal Navy adopted a less-than-adequate protection system proposed by Winn (1770) that consisted of a series of copper rods connected every few feet (every 60 cm or so) by connecting links. This chain of rods was attached to a rope hung from a metal spike at the top of the mast and loosely dangled into the sea. Neither the spike nor the down conductor was kept permanently in place. They were only rigged when storms were present or thought to be imminent. The Royal Navy's (Winn's) protection system had a number of serious drawbacks. Most important, the chain of rods was often not in place when lightning struck, and the system interfered with seamen working on the sails. In fact, three sailors were killed by lightning as they were erecting the linked conductors on an American ship-of-war in the Mississippi River (Tomlinson 1848). Additionally, arcing at the connecting links could melt the links, and no provision was made for bonding the protection system to nearby metal in order to avoid side flashes (see Section 4.4). Further, the captains of many ships believed that the protection system significantly increased the risk of a ship's being struck by lightning and hence were reluctant to put the protection in place when a storm approached (use of the system was at the captain's option). Despite the deficiencies noted above, the Winn system was the standard in the Royal Navy (and was widely used elsewhere) from about 1770 to about 1840.

In the 1770s, the French navy adopted a slightly improved version of the Royal Navy's lightning protection system. It used the same chain of rods but, rather than the chain dangling loosely from the spike at the top of the mast into the sea, the chain was routed down the permanent rigging and around the hull where it was connected to the ship's underwater copper sheathing for grounding. Copper sheathing was attached to the bottoms of most large ships of the era for protection in collisions with underwater objects and to impede marine borer worms. In the early 1800s, the chain conductors on French ships were replaced with metal cables.

A more adequate marine protection system than the original British or French systems, and one that contained most of the basic elements subsequently recommended in modern standards, was designed by William Snow Harris in 1820 and proposed to the British Admiralty in 1821 (Harris 1834, 1843). Harris was a physician turned electrical researcher whose interest in ships derived from his upbringing in Plymouth, one of the primary dock yards of the Royal Navy. Harris's system involved fixed (permanent) conducting plates that were routed along the mast down through the ship's hull to the copper sheathing on the bottom of the vessel. Harris recommended that all principal metallic masses in the hull be bonded to the lightning conductor in order to prevent side flashes. Harris spent about 25 years trying to convince the British Admiralty to adopt his system. It took (1) a successful field testing of the Harris system on eleven Royal Navy vessels starting in 1830, (2) an extensive campaign by Harris to publicize the extent of the previous lightning damage to the British Royal Navy, including a report listing the lightning-caused damage to 174 British naval vessels between 1793 and 1838, during which 62 deaths and 114 injuries occurred (Harris 1838, 1839), (3) the favorable recommendations of two study committees, and (4) administrative changes in the Admiralty (the powerful First Naval Lord who opposed Harris's system was replaced in late 1841, in a change of political administrations) before the British Royal Navy finally adopted the Harris system in June 1842. By 1850 all Royal Navy ships were equipped with the Harris lightning protection system.

Figure 10.1 shows an engraving illustrating the success of the Harris system in an 1846 lightning strike to HM Frigate *Fisgard*, which was at anchor on the Nisqually River in the Oregon Territory (the Territory encompassed the area of present-day

H. M. FRIGATE "FISGARD," PROTECTED BY HARRIS'S LIGHTNING CONDUCTORS FROM A SEVERE
STROKE OF LIGHTNING, SEPTEMBER 26, 1846.   *See p. 244.*

Fig. 10.1   An engraving of lightning striking HM Frigate *Fisgard* on September 26, 1846. The ship
suffered no significant damage by virtue of the lightning protection system of Sir William
Snow Harris. The engraving is one of several involving lightning by E. Whimper in the book
"*The Thunder-Storm*" by Charles Tomlinson (1848).

Oregon, Washington, and most of British Columbia). The drawing shows (1) two
lightning flashes to pine trees that are set on fire and (2) a bifurcated flash that
strikes the main mast both on its top and at a location a few feet above the deck, a
location where damage was found to the conductor of the lightning protection
system. At that location the copper plates were separated and "thrust asunder"
causing wood splinters from the mast to fall on the deck (Tomlinson 1848). The
damage just above the deck was probably due to a side flash (see Section 4.4) and
not to the separate lightning channel shown in the artist's interpretation of the
official damage report. Side flashes might well be expected in the relatively poor
"ground" of river water, as discussed in the next two sections. Tomlinson (1848)
reports that a side flash did occur to metal bolts "leading through the boatswain's
cabin" and "through the midshipmen's berth." Nevertheless, he states

Here is indisputable evidence, that as powerful a discharge of lightning as can be well
imagined, fell with force directly on the mainmast of the *Fisgard*, which expended all its
fury on the conductor, and was, by its protective influence, led securely to the sea, without
the slightest damage or inconvenience. We trace it from the points on which it first struck,
down to the very sea in which it finally vanished, and we find the ship unharmed and still
efficient amidst the blaze and crash of the most terrible element in nature.

The adoption of Harris's system by the British Navy in 1842 did not immediately lead to the wider adoption of permanently installed conductors. Chain conductors remained the standard in the United States, while many wooden merchant vessels (both British and other) continued to carry no protection at all. Not long after Harris succeeded in getting adequate lightning protection installed on all British Royal Navy ships, the wooden sailing ships of the Royal Navy and the navies of most other countries were replaced by iron steam ships. As noted at the beginning of this section, metal ships are inherently self-protected because the conducting metal of the ship provides a Faraday cage which, in terms of lightning protection standards, functions as the air terminal, the down conductors, and the means of grounding.

In 1831, Charles Darwin was preparing for his famous voyage to South America on *HMS Beagle*, a journey on which Darwin's observation of nature would lead him to propose the theory of evolution by natural selection. During the preparation for the voyage he attended a lecture by William Snow Harris on the subject of the lightning protection of ships. Darwin was accompanied to the lecture by the captain of the *Beagle*, Robert FitzRoy. Darwin wrote in his diary (Darwin 1933, p. 8–9)

Monday, November 21st: In the evening went to the Athenaeum & heard a popular lecture from Mr. [later Sir William Snow] Harris on his lightning conductors. By means of making an Electric machine a thunder cloud; a tub of water the sea; & a toy for a line of battle ship, he showed the whole process of it being struck by lightning & most satisfactorily proved how completely his plan protects the vessel from any bad consequences. This plan consists in having plates of copper folding over each other, let in the masts & yards & so connected to the water beneath. The principle from which these advantages are derived, owes its utility to the fact that the Electric fluid is weakened by being transmitted over a large surface to such an extent that no effects are perceived, even when the mast is struck by the lightning. The *Beagle* is fitted with conductors of this plan; it is very probable we shall be the means of trying & I hope proving the utility of its effects.

The *Beagle* was one of the 11 test ships specified by the Royal Navy for testing Harris's lightning protection system, in addition to the *Beagle*'s much better-known claim to fame. In the course of its most famous voyage, from 1831–1836, *Beagle* was involved in many thunderstorms and sustained no damage, but it is not clear if the ship was ever directly struck by lightning. On a later voyage, however, the *Beagle* was definitely struck without damage. Captain Sullivan describes a lightning strike to the *Beagle* at Monte Video, Uruguay, witnessed during his period of duty on deck (Tomlinson 1848).

Having been on board His Majesty's ship *Thetis* at Rio de Janeiro a few years since, when her fore-mast was entirely destroyed by lightning, my attention was always very particularly directed to approaching electrical storms, and especially on the occasion now alluded to, as the storm was unusually severe. The flashes succeeded each other in rapid succession, and were gradually approaching; and as I was watching aloft, the ship became apparently wrapt in a blaze of fire, accompanied by a simul-taneous crash, which was equal, if not superior, to the shock I felt in the *Thetis*. One of the electrical clouds by which we were surrounded has burst on the vessel, and as the mainmast at the instant appeared to be a mass of fire, I felt certain that the lightning had passed down the conductor on that mast. The vessel shook under the explosion, and an unusual tremulous motion could be distinctly felt. As soon as I had recovered from the surprise of the moment, I ran below to state what had happened, and to see if the conductors had been affected, when just as I entered the gunroom, Mr. Rowlett, the purser, ran out of his cabin (along the beam of which a main branch of the conductor passed), and said he was sure the lightning had passed along the conductor, for at the moment of the shock he heard a sound like rushing water along the beam. Not the slightest ill consequence was experienced, and I cannot refrain from expressing my conviction that, but for the conductor, the results would have been serious.

Interestingly, Darwin twice recorded in his diary the observation of St. Elmo's fire, a visible corona discharge on and near the surface of elevated masts in the thunderstorm's strong electric field (Darwin 1933, p. 80, p. 144).

Sunday, July 22nd. We are about 50 miles from Cape St. Mary's. I have just been on deck; – the night presents a most extraordinary spectacle; – the darkness of the sky is interrupted by the most vivid lightning. The tops of our masts & higher yard ends shone with the Electric fluid [Darwin annotates in a footnote: "St. Elmo's fire"] playing about them; the form of the vane might also be traced as if it had been rubbed with phosphorous.

Sunday, April 21st: At noon 300 miles from Maldonado, with a foul wind. Our usual alternation of a gale of wind & a fine day. We are off the mouth of the Plata. At night there was a great deal of lightning: if a hurricane had been coming, the sky could not have looked much more angry. Probably we shall hear there has been at M. Video a tremendous Pampero. Our Royal mast head shone with St. Elmo's fire & therefore, according to all good sailors, no ill luck followed.

In his diary, Darwin reports observing lightning damage to the main mast of a ship (not the *Beagle*), a house, and a church during the voyage (Darwin 1933, pp. 199–200). Darwin made an additional lightning-related observation in 1832 while the *Beagle* was moored in the La Plata River near Buenos Aires, on the southeastern coast of South America. In the nearby sand dunes he identified the first "fulgurite" specimens from the western hemisphere. Fulgurites are glassy, hollow tubes formed by lightning when it traverses poorly conducting (generally sandy) soil (see Section 12.3, Fig. 12.7 and 12.8). Darwin initially found small pieces of fulgurites on the sand surface and with further investigation discovered that they continued to greater depths. Subsequently, he excavated one fulgurite to a depth of two feet (Darwin 1897, pp. 59–62). Other fragments he recovered were apparently part of a fulgurite of length greater than 5 feet 3 inches.

10.2 Protective techniques

The placement of air terminals specified for boats in NFPA 780:2004 is determined using the rolling sphere approach with the sphere radius (striking distance) chosen to be 30 m (100 feet). That sphere radius, according to theory, assures protection from 97 percent of lightning flashes, all those with first stroke peak currents in excess of about 5 400 amperes (see Section 3.3, Section 4.2, and Table 4.1). An example of the rolling sphere method applied to the protection of a sailboat is illustrated in Fig. 10.2. Use of the essentially equivalent cone-of-protection method for a motorboat is shown in Fig. 10.3. Copper is the preferred conductor for marine

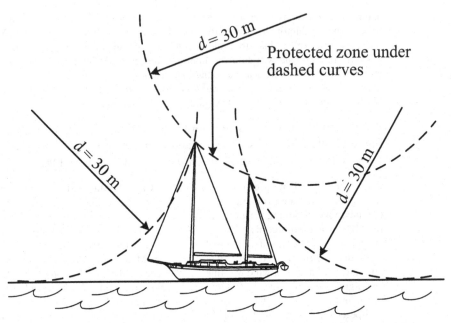

Fig. 10.2   A drawing adapted from NFPA 780:2004 of the design using the rolling sphere method of lightning protection for a sailboat with two masts (either conducting or outfitted with conducting downleads) with the taller mast in excess of 50 feet (15 m). Note that the top part of the front sail is left unprotected by the front mast. Reprinted with permission from NFPA 780, *Installation of Lightning Protection Systems*, Copyright ©2004, National Fire Protection Association, Quincy, MA 02169. This reprinted material is not the complete and official position of the National Fire Protection Association on the referenced subject which is represented only by the standard in its entirety.

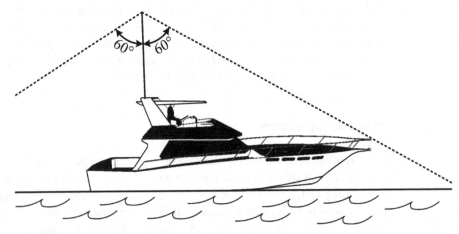

Fig. 10.3   A powerboat protected by a grounded vertical conducting mast that is assumed to provide a 60° cone of protection. The vertical mast, the air terminal, may be designed to be removable and used only when thunderstorms are present.

protection systems because of its resistance to corrosion, particularly in a salty atmosphere. Aluminum masts on sailboats are the air terminals and may be substituted for conventional down conductors which connect the air terminals to the grounding system. Aluminum masts are an exception to the recommendation for the use of copper. As is the case with other structural protection, junctions between aluminum and copper should be made with special stainless steel connectors in order to reduce potential corrosion at those junctions. Copper down conductors should be of a diameter equivalent or larger than No. 4 AWG (about 21 mm^2 cross-sectional area), unless there are multiple down leads. No. 6 AWG (about 13 mm^2 cross-sectional area) is deemed acceptable by NFPA 780:2004 for two separate down conductors in parallel. All metal objects on the boat should be bonded to the lightning protection system. A metal hull is the ideal grounding electrode. For boats with non-metallic hulls, conducting grounding plates or strips should be installed on the underside of the hull, beneath the waterline, to provide a path for the lightning current to flow into the water. Through-hull connectors should be metallic and have a cross-sectional area equivalent to a No. 4 AWG copper conductor. Under-hull grounding plates should be of minimum size 1 foot-squared × 3/16 inch thick (0.093 m^2 × 4.8 mm) according to NFPA 780:2004. As we shall discuss later, such grounding plates may provide a satisfactory ground connection in salt water, but they are probably not sufficient to inhibit side flashes above the water line in fresh water lakes and rivers. Thomson (1991) has shown that in fresh water, for an assumed water resistivity of 1000 ohm-meters (see Table 5.1), the grounding resistance of a one square foot grounding plate is 1500 ohms. Thus, a typical lightning current of 30 kA flowing into the water from a typically recommended 1 square foot grounding plate potentially results in a voltage difference of 45×10^6 volts (1500 ohms × 30 kA) between any metal rigging (and other connected metallic parts of a boat) and the water, potentially leading to destructive side flashes (see Section 4.4) to the boat and its occupants. On the other hand, unintended electrical discharges (arcs) from the grounding plate into the water below the water line may lower the grounding resistance, as discussed later, but certainly not to the point that hazardous voltages will not appear between the bonded conductors above water level and the surface of the water. The hazard is significantly less in salt water where the "ground" (water) resistivity is a thousand times lower and hence, above water, voltages will be in the tens of thousands of volt range for the example given above, insufficient to cause a spark of more than a few centimeters (an inch or less). Clearly, obtaining adequate grounding is the main obstacle to effective protection of boats and their occupants, as is also often the case in the protection of ground-based structures. Since controlling deleterious voltages is difficult on boats struck by lightning, all electronics should be protected by SPDs or other techniques (see Chapter 6).

One method of providing increased personal safety on a boat is to have available a small Faraday cage (see Section 3.1) into which the boat's occupants can retreat during a thunderstorm. The cage can be constructed of copper screen and should be bonded to the lightning protection system. If cleverly constructed, it can be folded up when not in use. Under no circumstances should boaters wait out a storm in the water near the boat (as is sometimes erroneously recommended). A lightning strike to the boat or to the water nearby could render the floaters unconscious and susceptible to drowning if they are not killed directly by lightning or by lightning-induced arcing from the boat.

Consistent with the availability of a lower "ground" resistance in salt water than in fresh water, there is a much lower occurrence of serious hull damage from side flashes in salt water than in fresh water, as is evident from Fig. 10.4 and Fig. 10.5, to be discussed in Section 10.3. In fact, since voltages developed in fresh water are relatively large, it is surprising that, in the study noted, 60 percent of protected boats in fresh water experienced no through-the-hull electrical breakdown. In this regard, the mitigating factor may be the dynamic ground resistance that arises when electrical breakdown occurs in the water as the lightning current flows out of the ground plate. This dynamic resistance is smaller than the dc resistance to uniform current flow because the effective area of the grounding electrode is enlarged owing to the volume of water rendered more conducting by the electrical discharges emanating from the grounding electrode. An analogous discussion for the behavior of grounding electrodes in soil is found in Section 5.4. Applying Eq. (5.10) to a hemispherical grounding conductor in fresh water with the assumption that the peak current I_p is 30 kA, the fresh water resistivity ρ is 1000 ohm-meters, and the breakdown electric field E_b is 1×10^6 V m^{-1}, we find that there will be breakdown arcs in the water to a radius r_{bd} of about 2 m. For a single ground plate, the effective area caused by the enlarging of the conducting area via arcing, according to Thomson (1991), is about 30 m^2 and is independent of the actual area of the plate. The effective resistance in fresh water is 72 ohms, and the maximum voltage for the assumed peak current is therefore 2.2×10^6 volts (72 ohms $\times$ 30 kA), a factor of about 20 smaller than that developed for a circular plate of 1 square foot (0.093 m^2) area. The voltage is, however, still large enough to make side flashes of meter length in air. The assumed value of 1×10^6 V m^{-1} for the breakdown electric field E_b in water in the example above is apparently a lower limit for a spherical electrode. A larger breakdown electric field for the same input current will result in a smaller r_{bd} via Eq. (5.10) and a higher maximum voltage than calculated above. The shape of the ground plate may also be important. A grounding conductor with sharp corners or points could initiate discharges at a lower voltage and result in a lowering of grounding resistance at a lower level of current than will a grounding conductor with a smooth surface.

The adequacy of the practice, endorsed by NFPA 780:2004, of directing the lightning current along a single down conductor (e.g., a sailboat mast) located in the middle of the boat, and then through the hull to a ground plate affixed to the bottom of the boat, is presently under discussion by the NFPA. In the protection of small, ground-based structures, the lightning current is generally routed to ground via down conductors located on the outside of the structure. The new scheme being considered by NFPA for boat protection is similar to the protection specified for small structures, involving directing the lightning current along the outside (rather than through the middle) of the boat to a bonding loop surrounding the boat at about deck level and from the bonding loop to the grounding electrode. The bonding loop additionally serves to equalize voltages on the boat within the loop, similar to the voltage equalization provided by a counterpoise surrounding a structure (see Section 5.1). When voltage differences are reduced, side flashes are suppressed.

10.3   Statistics

Thomson (1991) analyzed 71 reports of sailboat damage by lightning in Florida for which (1) the boat hull was constructed of fiberglass, (2) the mast was aluminum, and (3) there was clear evidence of lightning attachment to the top of the mast, usually in the form of damage to a masthead antenna. The resulting data were divided into four protection categories depending on whether the boats were in salt or fresh water and whether they did or did not have lightning protection systems when struck. A boat was considered to have a protection system if a connection existed between the base of the mast or shrouds and either a metallic keel or a ground plate below the hull.

The study found that about 3 percent of all moored sailboats in southwest Florida suffer lightning-induced damage to marine electronics each year. Bar graphs showing the frequency of occurrence of three degrees of electronics damage are given in Fig. 10.4. The frequency of occurrence is given as the percentage of the total number of boats in each of the four categories that fall in the particular damage class. In Fig. 10.4, the total number of boats in each of the four protection categories is: 26 with protection in salt water, 16 with no protection in salt water, 14 with protection in fresh water, and 11 with no protection in fresh water. Apparently, the present state of lightning protection is particularly ineffective for marine electronics. With or without lightning protection, boats in fresh water sustained more electronics damage to all systems than boats in salt water: 16 out of 25 (64 percent) in fresh water versus 18 out of 42 (43 percent) in salt water.

Figure 10.5 gives bar graphs showing the frequency of occurrence in each of the four protection categories that fall in each of the five levels of boat-hull damage. Damage to the boat hull was classified on a 0 to 4 severity index scale according to the following criteria: 0, no discernible burns or fractures; 1, small non-leaking cracks or burns; 2, small holes (typically described as "pin holes" of a millimeter or less diameter) that did not pose a threat of serious leaks; 3, large (several millimeter diameter) holes above the waterline; and 4, large holes (several millimeter diameter) below the waterline. Indices 2 to 4 represented breakdown through the hull. Boats with hull damage in category 4 were in sinking condition. In Fig. 10.5, the frequency of occurrence is given as the percentage of the total number of boats in each of the four protection categories that fall in each of the five levels of boat-hull damage. In Fig. 10.5, the total number of boats in each of the four protection categories is: 28 with protection in salt water, 16 with no protection in salt water, 15

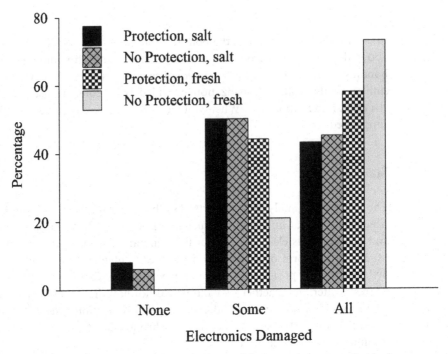

Fig. 10.4   Frequency distribution for proportion of boats with electronics systems that had none, some, or all of their electronics systems damaged as a result of a direct lightning strike. Adapted from Thomson (1991).

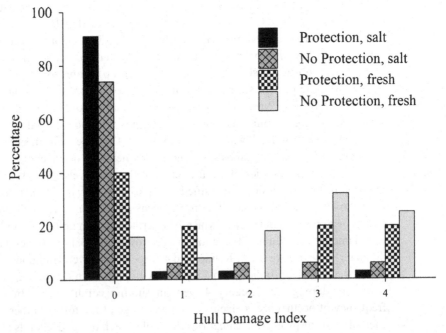

Fig. 10.5   Frequency distribution for the proportion of boats that incurred hull damage, on a 0 to 4 scale, as a result of a direct lightning strike. The index values are defined in the text. Adapted from Thomson (1991).

with protection in fresh water, and 12 with no protection in fresh water. A major difference is apparent in the severity of hull damage between boats struck in fresh and salt water. There is much more damage in fresh water as might be expected because of the poorer "grounding" conditions, as noted in Section 10.2. Hull damage of index 2 or higher, indicative of through-hull breakdown, certainly represents a failure of the lightning protection system.

References

Bernstein, T. and Reynolds, T. S. 1978. Protecting the Royal Navy from lightning: William Snow Harris and his struggle with the British Admiralty for fixed lightning conductors. *IEEE Trans. Education* **21**: 7–14.

Cerveny, R. S. 2005. Charles Darwin's meteorological observations aboard the H.M.S. Beagle. *Bull. Am. Meteorol. Soc.* **86**: 1295–1301.

Darwin, C. 1897. *Journal of Researches into the Natural History and Geology of the Countries Visited during the Voyage of the H.M.S.* Beagle *Round the World, under the Command of Capt. Fitzroy, R. N.* 2nd edn. D. Appleton and Company.

Darwin, C. 1933. *Charles Darwin's Diary of the Voyage of the H.M.S.* "Beagle". Cambridge: Cambridge University Press.

Harris, W. S. 1834. On the protection of ships from lightning. *Nautical Magazine* **3**: 151–156, 225–233, 353–358, 402–407, 477–484, 739–744, 781–787.

Harris, W. S. 1838, 1839. Illustrations of cases of damage by lightning in the British Navy. *Nautical Magazine*, enlarged series, **2**, 590–595, 747–748; **3**, 113–122.

Harris, W. S. 1843. *On the Nature of Thunderstorms.* London: John W. Parker, pp. 140–156.

NFPA 780:2004. *Standard for the Installation of Lightning Protection Systems.* Quincy, MA: National Fire Protection Association (NFPA).

Thomson, E. 1991. A critical assessment of the U.S. Code for lightning protection of boats. *IEEE Trans. Electromagn. Compat.* **33**: 132–138.

Tomlinson, C. 1848. *The Thunder-Storm.* London: The Committee of General Literature and Education, appointed by the Society for Promoting Christian Knowledge.

Watson, W. P. 1761. Some suggestions concerning the preventing the mischiefs which happen to ships and their masts by lightning. *Royal Soc. Phil. Trans.* **52**: 629–635.

Winn, J. L. 1770. A Letter to Dr. Benjamin Franklin, F. R. S. giving an account of the appearance of lightning on a conductor fixed from the summit of the mainmast of a ship, down to the water. *Phil. Trans.* **60**: 188–191.

11 Trees

11.1 General

Trees are protected in much the same way that structures are protected, with the added provision that lightning current must not be allowed to flow in the major root system. Current in the major tree roots can result in root damage that leads to the death of the tree. Grounding of the lightning current is therefore best accomplished by locating the primary grounding electrodes, ground rods and/or buried horizontal wires, near the outer edge of the branches where there are few significant roots. NFPA 780:2004 contains an Appendix, called Annex F, discussing the protection of trees, and the American National Standard Institute document, ANSI A300 (part 4: 2002), provides similar information and some additional information regarding the protection of specific types of trees.

According to NFPA 780:2004, all trees with trunks within 3 m (10 feet) of a structure or with branches that extend to a height above the structure should be protected. The three primary reasons for the NFPA-recommended protection are to avoid damaging side flashes (see Section 4.4) from the tree to the structure (the ground electrode of the tree is to be bonded to the structure protection system), to avoid a tree fire that could presumably spread to the structure, and to avoid the explosive and potentially damaging splintering of the tree due to the superheating of the moisture in the tree (see Section 11.3). ANSI A300, quoting the National Arborist Association, recommends the following additional candidates for protection: trees of historical interest, trees of unusual value, tall trees in recreation parks or park areas, and trees that are more likely to be struck by lightning owing to their location, such as isolated trees on hills, on golf courses, or in pastures. Clearly, tree protection serves both to save the tree from damage or death and to reduce the risk to humans, animals, and structures near the tree from potentially damaging side flashes and step voltages (see Sections 4.4, 5.3, and 7.4). Even if the tree is lightning-protected, these risks are not completely eliminated, so shelter should not be taken under a lightning-protected tree.

ANSI A300 assigns a level of susceptibility to lightning strike to 16 different types of trees when exposure of each is, in principle, the same. Listed as most susceptible are ash, tulip poplar, pine, oak, and hemlock. Listed as low in susceptibility are horse chestnut, beech, and holly. The basis for these assignments is not stated.

11.2   Protective techniques

An illustration of the lightning protection recommended by NFPA 780:2004 and ANSI A300 for a typical tree is found in Fig. 11.1. The protective system consists of metal wire or metal strip attached to the tree, extending from a well-established ground up the tree trunk to the top of the tree and out the major branches. The wire or strip serves as the down conductor. The top of this conductor can serve as the air terminal or a lightning rod may be attached there. Allowance must be made for wind sway and tree growth in attaching the wire or strip down conductors to the tree. The proper placement of the air terminals (or the tops of the down conductors) can be determined using the rolling sphere approach (see Section 3.3). According to NFPA 780:2004, grounding is best done as follows: extend three or more radial conductors in trenches 0.3 m (1 foot) deep, spaced at equal intervals about the tree base to a distance extending to the branch line but not less than 7.6 m (25 feet). Either install ground rods at the end of the radials or have the ends of the radial conductors bonded to a conductor that encircles the tree (ring electrode) at a depth of not less than 0.3 m. Bond the grounding system to any underground metallic water pipe or other significant metal body (such as a well casing) within 7.6 m of the branch line.

11.3   Types of damage

Lightning may strike a tree and leave it apparently unharmed (e.g., Orville 1968, Uman 1991), or it may cause structural damage to the tree without noticeable burning, or it may set the tree on fire. Photographs (among the best lightning photographs ever taken) of two unharmed trees are found in Fig. 11.2 and Fig. 11.3. The detailed effects of lightning on trees have been documented in a number of studies, most notably by Taylor (1964, 1965, 1969a,b) and by Schmitz and Taylor (1969). They found that most trees that were struck were not killed. The majority recovered from whatever lightning damage they might have sustained, although many were weakened and ultimately succumbed to attacks by insects and disease. Visible damage to tree trunks ranged from superficial bark flaking, to strip-like furrowing along the trunk, to virtually total destruction. Lightning damage that occurs to conifers (cone-bearing trees, mostly evergreens) and rough-barked deciduous trees such as oaks is shown in Figs. 11.4–11.6. Figure 11.4 shows a photograph of a Douglas fir tree that is shattered. Figure 11.5 illustrates the common spiral scars, strip-like furrowing, found on evergreens and rough-barked deciduous trees. A close up photograph of such a scar is shown in Fig. 11.6. Taylor (1964) examined 1000 lightning-damaged Douglas firs in western Montana. Most had shallow continuous scars a few inches wide along their trunks as shown in Fig. 11.5 and Fig. 11.6. About 20 percent had two or more scars (Fig. 11.5), 10 percent had severed tops, and about 1 percent had been reduced to slabs and

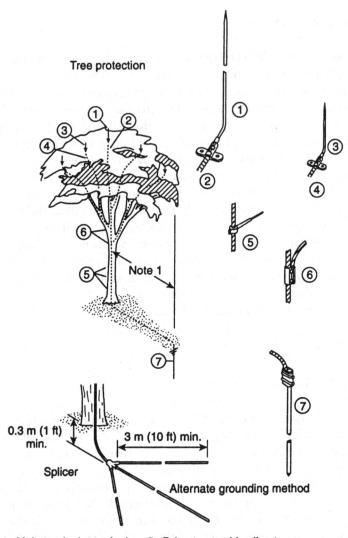

Tree protection

Note 1

0.3 m (1 ft) min.

3 m (10 ft) min.

Splicer

Alternate grounding method

1. Main trunk air terminal
2. Class I or Class II full-size cable
3. Branch air terminal
4. Secondary size cable
5. Drive-type cable clip at 0.9 m (3 ft) O/C
6. Splicer
7. Ground rod and clamp

Note 1: Locate ground approximately at branch line to avoid root damage.
Note 2: Install cable loosely to allow for tree growth.

Fig. 11.1   A drawing illustrating the lightning protection of a tree taken from NFPA 780:2004. Reprinted with permission from NFPA 780, *Installation of Lightning Protection Systems*, Copyright ©2004, National Fire Protection Association, Quincy, MA 02169. This reprinted material is not the complete and official position of the National Fire Protection Association on the referenced subject which is represented only by the standard in its entirety.

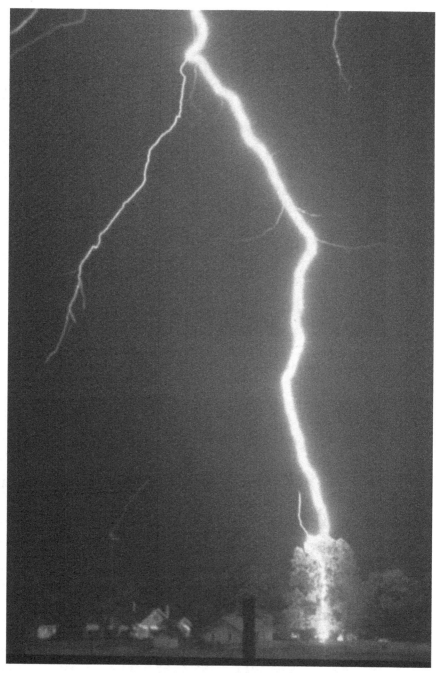

Fig. 11.2   A photograph of a 20-m-tall sycamore tree that was unharmed when struck by lightning. Discussion found in Uman (1991). Note the unconnected upward leaders from the struck tree and from the television antenna tower in the left of the photograph. The tower height was 21 m, and it was located 45 m from the tree. The unconnected upward leader from the tower was about 14 m in length, the one from the tree about 9 m. Photograph by Johnny Autery.

Fig. 11.3   A photograph of a 7-m-tall European ash tree that was unharmed when struck by lightning. Discussion is found in Orville (1968) and in Section 3.3. Photograph by R. E. Orville.

Fig. 11.4   Total destruction of a 70 foot Douglas fir in western Montana. Courtesy of the US Forest Service.

Fig. 11.5 Spiral scars on a Douglas fir in western Montana. Courtesy of the US Forest Service.

slivers (Fig. 11.4). Most of the scars on the Douglas firs were spiral; a few were straight. The average scar extended along 80 percent of the tree height, but none extended to the very tops of the trees. Scars either reached to ground level or close to ground level. Along the center line of the lightning scar there was often a crack which penetrated into the tree, as is evident in Fig. 11.6, and, when wood was

Fig. 11.6   Close up of a tree scar on a Douglas fir in Colorado. Photograph by Dr. F. Luiszer. Courtesy of D. G. Davis.

removed from a tree by lightning, it was usually ejected as two parallel slabs, separated along this crack. Figure 11.7 further illustrates the formation of the common tree scar. Sometimes, in place of the crack, lightning left a narrow strip of shredded inner-bark fiber fixed in a smooth shallow groove about 1/16 inch

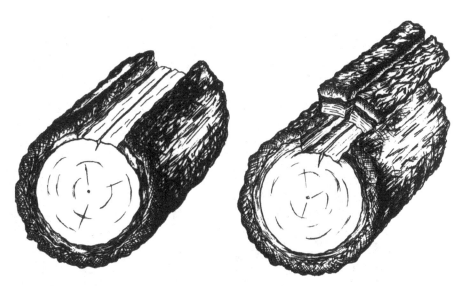

Fig. 11.7   Drawing of the formation of a lightning-caused tree scar as shown in Fig. 11.5 and Fig. 11.6. Courtesy of A. R. Taylor.

(0.16 cm) wide, as is evident in Fig. 11.6 showing the scar of a struck Douglas fir in Colorado. The relatively smooth barks of some deciduous trees, for example birches, allow for quite different damage characteristics. For these trees the bark is not removed in narrow, uniform strips but rather is torn off in large, somewhat irregular patches or sheets as shown in Fig. 11.8.

Sometimes a single lightning discharge can kill a group of trees. In a typical group kill, obvious lightning damage is visible on only one or two trees, often near the center of the dying group. As many as 160 trees have been reported killed this way, but in most cases the groups are probably smaller. It is unclear whether lightning does unseen damage to the roots of trees surrounding the struck tree or whether only the above-ground parts are affected by the discharge (Minko 1966). A perhaps-related phenomenon involving lightning and tree groups has been frequently documented by entomologists. They report that several species of bark beetles attack single trees damaged by lightning and then proceed to attack other trees surrounding the damaged one (e. g., Komarek 1964; Schmitz and Taylor 1969; Coulson *et al.* 1983, 1986; Schowalter *et al.* 1981; Lovelady *et al.* 1991). The result can be a group kill similar to those not involving bark beetles. Possibly, lightning is playing a hidden role in this type of group kill in that it may do unseen damage to the trees surrounding an obviously struck tree, thereby reducing their natural resistance to attack by bark beetles.

To set fire to a piece of wood, the ignition source must contact the wood for a sufficient period of time since initiation of burning depends both on a high applied temperature and on that temperature being present for some appreciable time. In general, the higher the temperature, the shorter the time. Moving your finger

Fig. 11.8   Lightning-caused damage to a smooth-barked birch tree. The bark is removed in large, irregular patches. Courtesy of the US Forest Service.

rapidly through a candle flame will not burn your finger, but leaving it in the flame for a few seconds certainly will. The lightning continuing current (see Section 1.3) provides a high enough temperature (roughly 10 000 °C or 20 000 °F) for a sufficient time (about 0.1 second) to ignite woody fuels. Fuquay *et al.* (1967, 1972) have shown conclusively that continuing currents in ground flashes are likely to cause forest fires. Since positive flashes generally, if not always, have a continuing current component that is associated with a relatively large charge transfer (see Section 1.3), positive flashes are prime candidates for setting forest fires (Fuquay *et al.* 1972). A tree may not necessarily be set on fire directly by the lightning current. Rather, the lightning current may ignite the leaves, moss, and other flammable material on the forest floor surrounding a tree, leading to the eventual ignition of the tree.

Studies of forest fuel ignition using laboratory discharges to simulate the lightning current have been performed by Latham and Schlieter (1989) and by Darveniza and Zhou (1994). There are so many variables in the fuel/discharge interaction that it is difficult to extrapolate the published laboratory results to the case of natural lightning.

While lightning without continuing current may or may not set fires, it can nevertheless be destructive. The bottom end of the lightning channel is often viewed as a current source which forces the lightning current into the struck object. This current generates heat in the object, the amount of heat depending on the object's ohmic resistance (see Section 2.3). If the object has a relatively high resistance (such as a tree) there can be a great deal of heating although not necessarily enough to cause burning. The rapid heating causes vaporization of some of the internal material. As a result, very high pressure is quickly generated within the material, and this pressure blows the material apart. For many of the tree scars examined by Taylor (1964), the lightning apparently followed a path through the cambium (a thin layer of living cells between the inner bark and the wood) or through the moist inner-bark tissue. These zones were apparently chosen as the lightning's paths because they offered lower resistance than the outer bark or the wood. The pressure generated in the cambial region expels a bark strip creating the scar shown in Figs. 11.5, 11.6, and 11.7. The pressure could also cause a split in the tree, but it is not clear why a strip of inner-bark fiber is pasted along the centerline of the scar, as is often observed. In some of the trees examined by Taylor (1964), wood as well as bark was blown out from the tree. In these cases the lightning current apparently traveled deeper within the trunk. Taylor (1964) found that older trees (over 200 years) were more likely to suffer wood-loss scars and suggested that perhaps the wood in old trees offers less electrical resistance to current flow than does the cambial zone.

Heidler *et al.* (2005) describe the severe destruction of two fir trees in Europe, both apparently by relatively large positive lightning flashes (see Section 1.3). In April 2000, the destruction of a fir occurred in a forest in the south of Germany about 100 km from Munich. The struck fir was about 32 m high and the diameter at the bottom was greater than 0.6 m. The fir splintered into three major fragments, each of which was estimated to have a weight of the order of half a metric ton. The explosion was so severe that major parts of the tree were blasted more than 10 m away from the remnant remains. Further, many smaller fragments weighing up to about 100 kg were found in the surrounding area at distances up to about 80 m from the tree. More than 10 surrounding trees had large areas of their bark damaged by fragments traveling with high speed. The German lightning locating system indicated that the fir was struck by a positive cloud-to-ground flash having a peak current of roughly 50 kA. A similar tree destruction was reported in Austria. According to the Austrian lightning locating system the tree was struck by a positive cloud-to-ground flash having a peak current of about 100 kA. While these lightning locating systems can provide an estimate of peak current, they do not provide information on the continuing current, total charge transfer, or action integral of the flashes, the latter parameter likely being most related to the damage that occurred (see Section 2.3). Lightning locating systems are discussed in Section 8.3.

11.4   The value of lightning to trees

Until recently, it was the general opinion that lightning-caused forest fires should be quickly suppressed, particularly those fires ignited near populated areas. Historically, lightning has played a significant role in shaping the nature of most forests via the fire it produces and via preferential destruction of the taller trees. Forests and lightning coexisted long before man was a significant feature of the landscape (e.g., Love 1970). Some trees, an example being the Douglas fir, have a very thick bark that insulates the tree against heat. Some pines have cones from which seeds can only be dispersed under the conditions of a wildfire. Wildfires clear both the forest floor and the canopy (allowing in more sunlight), producing optimal conditions for the establishment of seedlings. We may be indebted to ancient lightning-caused forest fires for the existence today of California's giant sequoias. The seedlings of these trees can germinate in ashes but are suppressed under the thick layer of needles that might cover an unburned forest floor. It is thought that in the southeastern United States, the balance between tall pines and shorter oaks has been maintained by the lightning destruction of the taller pines. If a sufficient number of pines had not been killed by lightning strikes (including the group kills discussed above), the pines would probably have shaded the oaks from the sun, significantly reducing the population of oaks. Before fire suppression became a general policy, frequent fires in the western United States kept the forest floor clean; the fires themselves were small and did little damage to the trees. Efforts to prevent and contain forest fires in the American West have enabled the brush to grow more thickly and now most fires are large. The number of record fires that have occurred recently (see Section 2.1) have enforced the view that regular prescribed burns and mechanical clearing of forest underbrush are essential to inhibit large fires. Prescribed burns serve the same purpose as lightning-caused fires and, in general, are easier to control. However, a prescribed burn got badly out of control near Los Alamos, New Mexico, in summer 2000, illustrating that the proper imitation of natural lightning is not a trivial pursuit. The disastrous Yellowstone fire of 1988 is discussed in Section 2.1, as is the issue of the cost of fighting forest fires and the indication that global warming is likely to continue to increase that cost.

A particularly interesting example of the interaction of lightning with vegetation is found in the unique tall grass prairie of the central United States. At one time, perhaps from about 10 000 years ago to about 100 years ago, Big Blue Stem, Indian Grass, and other tall grasses around 1.5 to 2 m high (5 to 6 feet) covered about half a million square miles of the Midwest. Over 99 percent of the original tall grass prairie is now farmland, and very fertile farmland at that, thanks to the previous history of the tall grass and lightning. Efforts are under way to regenerate some of the tall grass prairie since it is ideal for feeding cattle, as it once was for feeding herds of bison (a type of buffalo). To the north and to the south of the original tall grass prairie are forests, and to the west there is a short grass prairie. The fact that the region where the tall grass prairie was found did not support forests, when its rich soil would seemingly allow for the growth of forests, has long intrigued scientists. From an analysis of 100 years of climate records, Changnon et al. (2002) have suggested an answer to the question of why tall grass grew where apparently trees should have. The suggested answer is found in the unique weather of the region. The tall grass prairie had a greater percentage of drought years than did the forests to the north and south, and, additionally, the prairie had particularly dry winters. The final ingredient was fire, both fire from lightning and fire set intentionally by the Native Americans. When the tall grass prairie burned, the grass rapidly regenerated in the ash (fertilizer) of the burned grass, whereas any burned trees did not. Thus fire was essential to the maintenance of the grasslands that supported the Native Americans and the bison which they hunted. In fact, the Plains Indians were known to set fires intentionally in the tall grass so the new growth of grass would attract bison that the Indians could then kill more easily, or at least have to travel less distance to kill.

References

ANSI A300, part 4. 2002. National Arborist Association.

Changnon, S. A., Kunkel, K. E. and Winstanley, D. 2002. Climate factors that caused the unique tall grass prairie in the central United States. *Phys. Geog.* **23**: 259–280.

Coulson, R. N., Hennier, P. B., Flamm, R. O. *et al.* 1983. The role of lightning in the epidemiology of the southern pine beetle. *Z. Angew. Entomol.* **96**: 182–193.

Coulson, R. N., Flamm, R. O., Pulley, P. E. *et al.* 1986. Response of the southern pine bark beetle guild to host disturbance. *Environ. Entomol.* **15**: 859–868.

Darveniza, M. and Zhou, Y. 1994. Lightning-initiated fires: energy absorbed by fibrous materials from impulse current arcs. *J. Geophys. Res.* **99**: 10663–10670.

Fuquay, D. M., Baughman, R. G., Taylor, A. R. and Hawe, R. G. 1967. Characteristics of seven lightning discharges that caused forest fires. *J. Geophys. Res.* **72**: 6371–6373.

Fuquay, D. M., Taylor, A. R., Hawe, R. G. and Schmidt, C. W. Jr. 1972. Lightning discharges that have caused forest fires. *J. Geophys. Res.* **77**: 2156–2158.

Heidler, F., Diendorfer, G. and Zischank, W. 2005. Examples of trees severely destructed by lightning. *International Conference on Lightning and Static Electricity*, ICOLSE 2005, September 19–23, 2005, Seattle, WA. Paper GND-22.

Komarek, E. V. Sr. 1964. The natural history of lightning. *Proc. 3rd Annual Tall Timbers Fire Ecol. Conf.* 139–186.

Latham, D. J. and Schlieter, J. A. 1989. Ignition probability of wildland fuels based on simulated lightning discharges. *USDA Forest Service Research Paper* INT-411.

Love, R. M. 1970. The rangelands of the western United States. *Sci. Am.* **222**: 88–96.

Lovelady, C. N., Pulley, P. E., Coulson, R. N. and Flamm, R. O. 1991. Relation of lightning to herbivory by the southern pine bark beetle guild (Coleoptera: Scolytidae). *Environ. Entomol.* **20**: 1279–1284.

Minko, G. 1966. Lightning in radiata pine stands in northeastern Victoria. *Australian Forester* **30**: 257–267.

NFPA 780 (National Fire Protection Association). 2004. *Standard for the Installation of Lightning Protection Systems.* Quincy, MA: National Fire Protection Association.

Orville, R. E. 1968. Photograph of a close lightning flash. *Science* **162**: 666–667.

Schmitz, R. F. and Taylor, A. R. 1969. An instance of lightning damage and infestation of ponderosa pines by the pine engraver beetle in Montana. *USDA Forest Serv. Res.* Note INT-88.

Schowalter, T. D., Coulson, R. N. and Crossley, D. A. Jr. 1981. The role of southern pine beetle and fire in maintenance of structure and function of southeastern coniferous forests. *Environ. Entomol.* **10**: 821–825.

Taylor, A. R. 1964. Lightning damage to trees in Montana. *Weatherwise* **17**: 62–65.

Taylor, A. R. 1965. Diameter of lightning as indicated by tree scars. *J. Geophys. Res.* **70**: 5693–5695.

Taylor, A. R. 1969a. Lightning effects on the forest complex. *Proc. 9th Annual Tall Timbers Fire Ecol. Conf.*, 127–150.

Taylor, A. R. 1969b. Tree-bole ignition in superimposed lightning scars. *USDA Forest Serv. Res.* Note INT-90.

Uman, M. A. 1991. The best lightning picture I've ever seen. *Weatherwise* **44**: 8–9.

12 Overhead and underground power and communication lines

12.1   Overview

Both overhead (above-ground) and underground power and communication lines are susceptible to lightning-caused interruption and permanent damage. In an open field, the probability of lightning striking a buried cable is less than, but not that much different from, the probability of its striking an overhead line, as we shall discuss in Section 12.3. If a lightning flash would randomly strike the ground in the "near vicinity" of the location of either an underground or an overhead line in the case where the line were not present, that lightning would likely strike the line when the line was present. "Near vicinity" is typically a horizontal distance of some tens of meters. The attraction of lightning to objects projecting above ground level has been discussed in Section 1.5 and Section 3.3 and will be further considered in this section.

The number of strikes to an overhead power line per kilometer of line per year, N_{km}, can be roughly calculated by using the concept of an "equivalent collective area" on the ground in which the lightning would strike if the line were not there (see Section 1.5). We assume that all flashes that would have hit ground within a horizontal distance equal to two line heights on either side of the line center (if the line were not there) will strike the line. The resulting equivalent collective area is shown shaded in Fig. 12.1. For a length of line ℓ, that area is $4h\ell$, where h is the line height. Multiplication of the area $4h\ell$ in square kilometers by the ground flash density N_g in ground flashes per square kilometer per year ($km^{-2}yr^{-1}$) yields the number of strikes per year to a length ℓ of line. For a one kilometer length of the line, ℓ is set equal to 1 km, and

$$N_{km} = 4hN_g \ km^{-1}yr^{-1} \quad (h \text{ expressed in kilometers}) \tag{12.1a}$$

or

$$N_{km} = 0.004hN_g \ km^{-1}yr^{-1} \quad (h \text{ expressed in meters}) \tag{12.1b}$$

Thus, in a good portion of Florida where $N_g = 10 \, km^{-2}yr^{-1}$, a 10-m-high line (a typical distribution line), will suffer 0.4 strikes per kilometer of line per year or one strike every 2.5 kilometers each year (roughly a strike each 1.5 miles per year). In the northeastern United States, where N_g is roughly one-fifth the value in Florida (see Fig. 1.5), a 10-m-high power line will be struck one-fifth as often as in Florida, or once

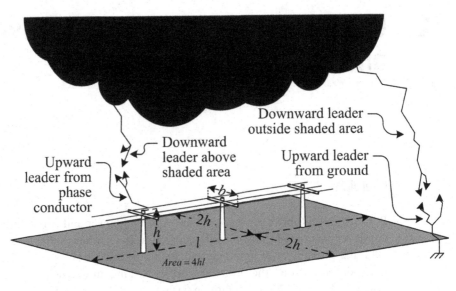

Fig. 12.1   A drawing illustrating the calculation of the number of lightning flashes striking an overhead power line. The equivalent collective area is shown shaded.

every 12 or 13 kilometers of line each year. These calculations can probably be considered accurate to a factor of two or so, the primary unknown in the calculation being the horizontal distance that the power line can send an upward and outward-moving leader to connect with a downward-moving stepped leader, here assumed to be twice the line height. There are more complex expressions than Eq. (12.1) found in the literature from which to calculate the strike rate per line length, but it is not clear that any expression is more accurate. There is relatively little experimental data to validate any given expression. For example, Eriksson (1987) has proposed the formula

$$N_{\mathrm{km}} = 0.001 N_{\mathrm{g}} \left(28h^{0.6} + b\right) km^{-1} yr^{-1} \tag{12.2}$$

where b is the width of the distribution line structure (see Fig. 12.1), with h and b expressed in meters. For $h = 10\,\mathrm{m}$, $b = 1\,\mathrm{m}$, $N_{\mathrm{g}} = 10\,\mathrm{km}^{-2}\,\mathrm{yr}^{-1}$, Eq. (12.2) yields a value roughly twice that of Eq. (12.1). Additionally, the rolling sphere method with an assumed sphere radius (striking distance) can be used to define the equivalent collective area (see Sections 3.3, 3.4, and 4.2). Note that nearby buildings and trees may play a role in the lightning performance of distribution lines by intercepting lightning flashes that otherwise would have directly hit a line, decreasing the number calculated from any expression like Eq. (12.1) or (12.2).

12.2   Overhead power lines

Protection of overhead electric power lines from the deleterious effects of lightning may be achieved by one or a combination of the following methods: (1) use of

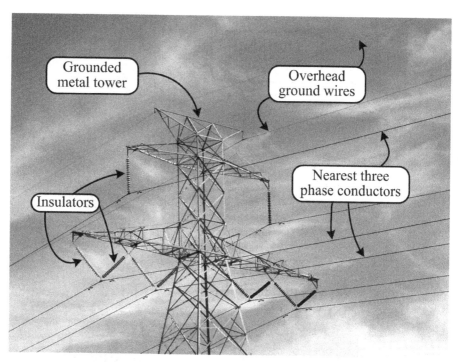

Fig. 12.2   A transmission line in central Florida with two sets of phase conductors and two overhead
ground wires attached to the metal transmission line tower. Photograph by Keith Rambo.

the highest economically reasonable insulation levels, (2) use of overhead ground
wires with good connections to Earth at the most closely spaced intervals physically
possible and economically reasonable (see Figs. 1.10, 12.2, 12.3), and (3) use of
arresters between the phase conductors and the neutral, spaced as closely along
the line as economically reasonable and at locations of sensitive line hardware
(see Fig. 12.4).

Either a direct lightning strike on a power line or an induced voltage from a
nearby strike may lead to line "flashover" (electrical breakdown between the wires
of the line), and/or failure of arresters, transformers, insulators, or other line
hardware. Flashovers or equipment failure can result in an out-of-service line,
an "outage." A photograph of a line flashover is found in Fig. 2.3. Direct strikes
are the most difficult to protect against because the associated overvoltages can
be many millions of volts. As noted in Section 2.3 and further discussed in
Section 12.2.2, a typical return-stroke peak current of 30 kA (where 15 kA flows
in each direction from the strike point) multiplied by the parallel combination of
the typical line surge impedance "seen" in each direction (about 500 ohms) yields a
voltage at the strike point of 7.5×10^6 volts. On distribution lines the line insulation
can generally withstand only between 100 and 300 kV. Voltages induced on overhead lines by nearby strikes are thought to be less than about 300 kV, so induced
voltages are easier to protect against, although there are no reliable data on the

Overhead
ground wire

Stand-offs

Ground lead

Insulators

Phase
conductors

Fig. 12.3   An overhead ground wire protecting the phase wires of a wood-pole distribution line in Florida. The down conductor is routed into the air around the vertically configured phase conductors to provide greater insulation against a flashover between the overhead ground wire/connected grounding system and the phase conductors. Photograph by Jens Schoene.

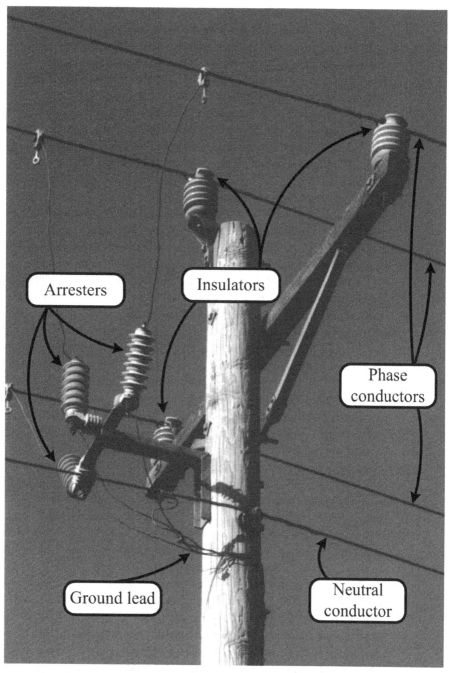

Fig. 12.4   A wood-pole distribution line in Florida that is protected by arresters connected between each of the three horizontally configured phases and ground.

fraction of line flashovers due to nearby strikes (there are relatively many nearby strikes) vs. direct strikes (of which there are relatively few; see Section 12.1).

In general, it should not be possible for nearby lightning to strike ground within a few tens of meters of an overhead power line. Rather, it will strike the line directly. The exception is the case where there are trees or structures very close to a line which can be preferentially struck, providing a very close lightning current that can induce a relatively high voltage on the line. Lightning protection against induced effects is basically the same as for direct stroke protection, so protection for the latter generally obviates the need to protect specifically for the former. The calculation of voltages induced on overhead lines by nearby lightning is considered by Borghetti *et al.* (2007), Nucci (1995), Nucci and Rachidi (1995), and Agrawal *et al.* (1980), among others.

Lightning protection for the two general classes of power lines, transmission lines and distribution lines, is generally considered separately by utilities, although the basic principles of protection are the same for both. Transmission lines transport electric power cross-country and operate at relatively high voltage levels, from near 50 kV to near 1000 kV (1 MV). Because transmission lines carry high voltages, they necessarily must have high insulation levels (to avoid electric breakdown between the lines due to the high operating voltages). To achieve this high insulation level the individual wires are placed far apart and are suspended on large insulators, as can be seen in Fig. 1.10 and Fig. 12.2. The inherently high insulation levels for transmission lines render them less susceptible to lightning damage than distribution lines, and hence they suffer fewer outages than do the lower voltage, less-well-insulated distribution lines, although the consequences of a transmission line failure can be much more costly (see Section 2.1). Distribution lines typically operate at voltages near 10 kV, with a range from near 2 kV to near 50 kV, and distribute electric power within cities and particularly to transformers near commercial and residential structures where the distribution line voltage is converted to the low voltage (e.g., 120 and 240 volts) used to power electric lights, electronics, and motors. Very often, the wires of a distribution power line serve as the lightning protection for telephone and cable TV lines mounted beneath the power lines. Similarly, distribution lines are sometimes affixed to the lower portion of the concrete, metal, or wood poles of a transmission line and hence are protected from a direct lightning strike by the transmission line.

The phase conductors of a power line carry the line voltage and are insulated from each other, from the "neutral" conductor, and from ground. Generally, the neutral conductor is attached to grounding electrodes via periodic down conductors. The neutral may be located either underneath the phase conductors or above them. A location underneath provides some protection against accidental contact of the phase conductors by, for example, metal ladders carried upright. An overhead location for the neutral or other overhead wire grounded at multiple points along the line provides lightning protection similar to the protection of structures involving standard lightning protection: air terminals on top of the structure, downleads, and grounding. Overhead ground wires protecting power lines are also similar to launch vehicle protection by the long overhead wires grounded at each end shown in Fig. 4.1 and Fig. 4.2. We now discuss in more detail the lightning protection afforded by overhead ground wires, followed by a discussion of the lightning protection provided by lightning arresters and circuit breakers.

12.2.1   Overhead ground wires

Grounded wires located above the phase conductors are referred to by a variety of names: overhead ground wires, shield wires, and sky wires. Overhead ground wires are most commonly used for the lightning protection of transmission lines where they are typically attached to the metal towers supporting the lines (Fig. 1.10 and Fig. 12.2). Those conducting towers serve as down conductors for the lightning current. Overhead ground wires are not commonly used on distribution lines except in regions of relatively high lightning activity. On distribution lines the down conductors are generally bare wires affixed to the wood poles of the line. If the phase conductors are mounted vertically, as shown in Fig. 12.3, the down conductors ("ground lead" in Fig. 12.3) should be routed through the air around the phase conductors on their path to the grounding electrode to provide greater insulation between the phase wires and ground. The tops of wood poles on distribution lines are often fitted with a small air terminal or are covered with a metal cap. The intent is to keep lightning current from entering the top of the wood pole and splitting or otherwise damaging the pole. When concrete poles are used to support power lines, there is usually a down conductor or grounding wire inside the pole (placed there during construction of the pole) that exits at the upper and lower sections of the pole. The internal wire is bonded to any reinforcing steel in the pole.

While overhead ground wires are generally effective in providing protection for the phase conductors against the effects of lightning, they do exhibit two failure modes. (1) Lightning current injected into an overhead ground wire can result in "back-flashover" from the ground wire to a phase wire if the resultant voltage on the ground wire is high enough, as can occur if there is the combination of a large lightning current and inadequate grounding of the overhead ground wire. (2) Lightning may bypass an overhead ground wire and strike a phase wire directly if the overhead ground wire is not properly located, and sometimes even if it is. The "shielding angle" is defined as the angle between an imaginary vertical line extending downward from the overhead ground wire and an imaginary line connecting the ground wire and phase conductor, as shown in Fig. 12.5. The shielding angle for the line shown in Fig. 1.10 is nearly zero degrees, a desirable situation. Shielding angles of about 30 degrees have been used with relatively good success on the great majority of transmission lines with tower heights in the range of 25 m, as in Fig. 12.2. Taller lines require smaller shielding angles than 30 degrees or else there will be "shielding failures"; that is, the lightning will bypass the overhead ground wire and strike the phase wire, consistent with the predictions of the electrogeometric and rolling sphere models (see Sections 3.3 and 3.4). Even well-located overhead ground wires will fail to intercept some flashes since downward-moving

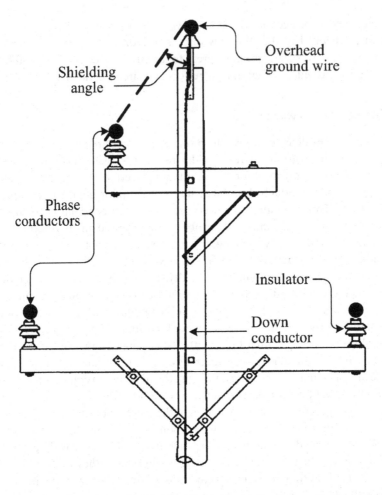

Fig. 12.5   An illustration of the "shielding angle" that the overhead ground wire provides for the phase conductors of a power line. Adapted from IEEE Standard 1410:2004. Copyright © 2004 IEEE. All rights reserved.

leaders with relatively small charge densities (leading to relatively small return strokes) can bypass the ground wire without inducing an upward connecting leader from the ground wire, but can induce a connecting leader from the phase wire beneath, at least according to the electrogeometric and rolling sphere models. The situation is similar to the case of a lightning strike to the side of a tall building, where the top of the building is bypassed by the downward-moving stepped leader (see Section 3.4 and Fig. 3.10). For the case of small-current strokes contacting a transmission line phase wire directly, flashover of the line insulation may not occur because such small stroke current in the phase conductor may not produce a large enough phase conductor voltage to lead to flashover for the high insulation level of the transmission line. The primary problem in designing transmission line protection, therefore, involves identifying an appropriate position for the overhead ground wires so that they intercept strokes having prospective peak currents above some minimum amplitude.

The electrogeometric model (see Section 3.3) has been successfully used not only to determine the placement of the overhead ground wires on lines of different heights and geometries, but also to model the long-term outage rate for existing lines and to predict the outage rate for lines not yet constructed via a Monte Carlo approach (e.g., Sargent and Darveniza 1967, 1970; Currie *et al.* 1971; Anderson 1981; and Liew and Darveniza 1971, 1982a,b). In these Monte Carlo calculations, the peak current (striking distance), angle of leader approach to the line, leader distance from the line, and perhaps other variables are chosen randomly, over and over again, within the range of measured or assumed lightning characteristics. Given the constraint of the known ground flash density and given the known geometrical configuration of the line, one can use a computer to calculate in a relatively short time the effects of many years' worth of random strike data, thus determining long-term averages of annual strikes to the overhead ground wire, shielding failures, and outage rates.

Finally, while we are discussing the electrogeometric model, Fig. 12.6 illustrates the use of its simplest incarnation, the rolling sphere method (see Section 3.4), to determine the placement and height of conducting vertical masts that protect an electric power substation from direct lightning. The vertical masts are typically bonded to a buried ground grid, and sometimes overhead ground wires are strung between the mast tops. These shield wires can be a continuation of the overhead ground wire or wires protecting the power lines entering and exiting the substation.

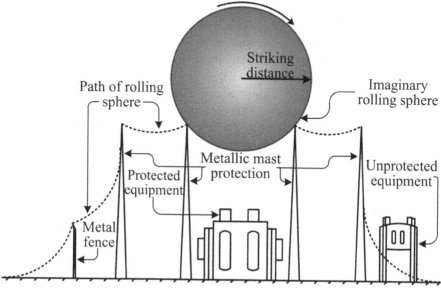

Fig. 12.6   Design of lightning protection for an electric power substation using metallic masts bonded to a grounding mesh and the rolling sphere method. Note that an unprotected transformer is shown. Adapted from IEEE Standard 998:1996. Copyright © 2006 IEEE. All rights reserved.

12.2.2   Arresters and circuit breakers

We consider now the application of surge protective devices (SPDs), most often referred to as "lightning arresters" when used for power lines protection; and further, when all other lightning protection has failed, the use of circuit breakers to reduce or eliminate the effects of that failure. General aspects of SPDs are discussed in Chapter 6. Lightning arresters serve to clamp the lightning-caused voltages between phase and neutral wires to values below that which would cause a flashover through the air between those wires or would damage hardware mounted on the line such as pole-mounted step-down transformers. Arresters protecting the three phases of a distribution line from flashover are shown in Fig. 12.4. If such a flashover does occur, the 60 Hz power current (called "power-follow" current) may flow through the lightning flashover path resulting in the long-duration short-circuiting of the lines, which can in turn result in the power-follow current cutting through one or more of the lines ("burn down" of the lines) by the mechanism discussed in Section 2.3, Eq. (2.1). To deter this more serious lightning outage, a circuit breaker located in the substation that feeds the line (or a similar device called a "recloser" located on the local distribution line circuit) senses the short-circuit current and interrupts the power current flowing in the line by physically making a break in the line, thus turning off the power frequency (60 Hz) voltage on the line and extinguishing the power-follow air discharge. When open, the circuit breaker also interrupts all electricity flow from the power line to its customers. After about a second, when the lightning is presumed to have ended, the circuit breaker auto-matically closes (re-connects the break in the line it had previously created), ideally re-powering the line on which direct lightning current and lightning-caused power-follow current no longer flow. While arrestors are very useful in limiting lightning-caused distribution and transmission line transient voltages, they have the potential disadvantage of permanently short-circuiting the lines to ground in the event of an arrester failure. The line will then be out of service (suffer an outage) until a repair crew arrives. Such a short-circuit failure, generally of the arrester's MOV blocks (most arresters now used are of the MOV type – see Section 6.4), will likely occur on a typical distribution line if a large enough lightning stroke contacts the line very close to the arrester. From triggered lightning tests (see Section 13.2) and theory, it has been determined that the energy input from natural first stroke waveforms with peak currents two to three times greater than the median of about 30 kA will damage the MOV blocks of the two distribution arresters closest to the strike-point (McDermott 2006, Schoene *et al.* 2007a,b). During the first 100 µs or so of the return stroke current, almost all the return stroke current passes through the closest arrester on either side of the strike-point. This is the case because the inductance of the line sections between the nearest two arresters and the next closest arresters provides a relatively high impedance to the rapidly varying initial return stroke current compared with the lower impedance of the nearest arresters. The situation is similar to that of the multi-stage SPD protection circuits discussed in Section 6.6 where individual SPDs are separated electrically by decoupling impedances. Following the first 100 μs or so, after the return stroke current variation is no longer rapid, the lightning current spreads out among the many arresters on the line, minimizing the energy input to any single arrester. The occurrence of subsequent strokes (strokes following the first stroke) in a flash will increase the likelihood of closest arrester failure by depositing additional energy in the MOV blocks of the closest arresters. Continuing current may deposit even more energy, but this slowly varying current will generally be shared by the multiple arresters on the line and be shunted to ground through line and substation transformers. Thus, its effect is not expected to be as important as the effect of the return strokes.

In order to prevent an arrester-induced short-circuiting of a distribution or transmission line, arresters are often outfitted with "ground lead disconnectors." Ground lead disconnectors are used commonly in the United States but are not permitted in some countries, apparently because the disconnectors are activated by explosive devices. The disconnectors operate when the heat from the 60 Hz short-circuit current (also called "fault" current) causes an explosive cartridge (usually a filled 22 caliber bullet casing) to propel the arrester ground lead downward and away from the body of the arrester, thus interrupting the fault current by rapidly disconnecting the arrester from the circuit. Disconnectors are designed to operate in a time of 5 ms or less for relatively high fault currents and in a few tenths of a second for low fault currents. Representative low and high fault currents on distribution lines are 20 amperes and some kiloamperes respectively, and on transmission lines, 1 kA and some tens of kiloamperes, respectively. Disconnectors are designed so that the energy associated with lightning should not cause them to fire, but they should operate on the energy input from the 60 Hz fault current that passes through the arrester in the event of MOV block failure caused by lightning. Information on the testing standards for distribution arresters is found in IEEE Standard C62.11:2005.

Fuses are also occasionally used to interrupt 60 Hz fault current flowing through arresters or through other electrical equipment or circuits, disconnecting them from the energized line until a repair crew arrives. Lightning currents should generally not provide sufficient energy to operate such fuses.

The degree of protection provided by differing arrester spacings (every pole, every other pole, every third pole, etc.) is poorly understood. Placing an arrester at every pole can apparently provide reasonably good protection against lightning-caused flashovers, although there can be arrester failures if the lightning imparts a much higher current or energy to an individual arrester than specified by the standard to which the arrester was manufactured and tested, as discussed above. The downside to using a relatively small spacing between line arresters is the initial expense of additional arresters and increased maintenance to check and replace the arresters when they fail. The compromise generally adopted is to place arresters every third or fourth pole and at sensitive locations such as transformer stations and underground cable entrances. There is simple theory available to predict failure rates for different arrester placement (e.g., IEEE Standard 1410:2004) but little experimental data to allow a well-informed decision to be made about spacing. The physical situation can be described by way of example. Assume there are arresters every other pole with 75 m between poles, and let lightning with current $I(t)$ strike midway between two arrester poles, that is, 75 m distant from each. A voltage wave

$$V(t) = \frac{1}{2} Z_0 I(t) \tag{12.3}$$

where Z_0 is the characteristic impedance of the line, typically near 500 ohms, is both impressed at the strike-point and propagates away from the strike-point in each direction. The voltage is impressed between the struck line and distant ground, but the voltages between the struck phase and the other phases, and between the struck phase and the neutral, are not too much different from the voltage between the struck phase and ground, perhaps 10 to 30 percent less. As noted earlier and expressed in Eq. (12.3), the voltage at the strike-point will increase to many millions of volts as the injected current rises to its peak value if relief is not provided by the arresters. For subsequent strokes, the current rise time is tenths of a microsecond to about a microsecond (see Fig. 2.2); for first strokes the current waveform exhibits a few microseconds of current increase to about half of peak value followed by a fast transition to peak similar in character to the total rising portion of the subsequent stroke current waveform (see Fig. 2.1). When the two outward-propagating voltage waves, which have the same wave shape as the causative current, contact the arresters, those arresters act as imperfect short circuits: they clamp the voltage between the arrester and the neutral at some tens of thousands of volts and reflect back toward the current injection point a reverse-polarity version of the incoming voltage wave. When the reverse-polarity voltage wave arrives at the injection point, it reduces the voltage there to a value near the arrester clamping value. Any line flashover that is to occur must occur before the relief voltage wave arrives, in this case about 0.5 μs (a 150 m round trip at the speed of light, $3 \times 10^8\,\mathrm{m\,s^{-1}}$, although the speed of propagation will be slightly slower than the speed of light on a typical line so the travel time will be a bit longer). Thus, for current rise times near 0.5 μs a flashover at the current injection point could well occur before the relief voltage wave arrives, depending on the time it takes for the flashover to bridge the gap between the struck phase and other conductors. The time for a flashover to develop in the air between the struck phase and one of the other phases or the neutral will depend on the applied voltage (the larger the voltage, the shorter the time) and on the size of the air gap, but generally will be near or less than 1 μs. Thus, for the example given above, the system is certainly on the verge of flashover, and perhaps flashover is likely unless the lightning currents have relatively slow rise times and the line is very well insulated. Clearly, placing an arrester on every pole, and thereby reducing the arrival time of the relief voltage wave from the example given above, would provide the best (and most expensive) protection against flashovers and potential outages.

Schoene *et al.* (2007a,b) have injected triggered-lightning currents (see Section 13.2) into two distribution power lines with about 50 to 60 m span lengths, one with arresters 1.5 spans from the current injection point in either direction, and the other with arresters 1.5 spans away in one direction and 2.5 spans in the other direction. The latter line flashed over during 90 percent of 82 injected return stroke currents, the former during about 25 percent of 34 stroke currents. The latter exhibited a time of 0.7 μs until the relief wave arrived at all points on the line, while the former needed only 0.5 μs. The rise times for return stroke current were between about 0.7 and 2 μs in the latter case and were not properly measured for the former but were probably similar. The mean return stroke currents were near 15 kA in both cases, about half the typical value for natural first strokes. The mean value of peak current for the triggered strokes that caused flashover was greater than the mean value for those that did not cause flashover, but relatively small peak currents sometimes caused flashovers.

Theory from IEEE Standard 1410–2004, illustrated in its Fig. 10, which is somewhat different from the theory presented above, predicts that for a distance of 150 m between arresters (two 75 m spans) and a critical flashover voltage (CFO) of a line of 350 kV, about 70 percent of first stroke currents will cause flashovers, with that percentage increasing as the distance between arresters is increased, but that there will be no flashovers if there is an arrester every 75 m. The assumptions in this calculation include a current rise time of 2 μs. For a CFO of 150 kV and the same 150 m arrester spacing, the IEEE Standard calculations predict a flashover for every first stroke.

12.3   Underground cables

Underground power and communication lines are almost always protected by SPDs. Often underground communication lines, typically buried about 1 m (about 3 feet) deep, have a bare "ground" wire or wires 0.3 m or so above them in direct contact with the soil to intercept and dissipate the lightning current. Such a buried ground wire is sometimes called a counterpoise. In the undesirable event of a flashover from a counterpoise to the power or communication line below it, the current to that line will at least be significantly reduced from the value that a direct strike would have delivered.

When lightning strikes the Earth, the strike-point on the ground can be considered to be a small volume of ground of characteristic dimension r_0 about 0.1 m or less. From the discussion in Section 5.4, and the re-arrangement of Eq. (5.4) leading to Eq. (5.10), we can show that lightning current will cause the breakdown electric field strength in the ground E_b to be exceeded to a radius

$$r_{bd} = \sqrt{\frac{\rho I_p}{2\pi E_b}}, \quad r_{bd} > r_0 \qquad\qquad (5.10), (12.4)$$

assuming a hemispherical strike region. From Eq. (12.4) with a typical breakdown field in the ground $E_b = 300\,\text{kV}\,\text{m}^{-1}$ (see Section 5.4), a soil resistivity of

1000 ohm-meters (Table 5.1), and a typical peak current of 30 kA, the hemispherical breakdown radius is about 4 m. For non-uniform arc formation, longer arcs will occur. Figures 5.5 and 5.6 show the results of surface arcing in the laboratory and on a golf course green, respectively. Presumably, there was also underground arcing. Lightning-caused furrows in the soil of many tens of meters length are not uncommon. If there is buried metal in the vicinity of a strike to ground, such as an underground cable, the non-uniformity of the electric fields underground will serve to attract the arc to the conductor over a longer path than would otherwise occur.

Underground lightning can melt and convert the poorly conducting soil to ionized gas, not unlike the case of the lightning channel in air. When sandy soil is so heated and subsequently cools, a fulgurite, a hollow glassy tube, may be formed, leaving a record of the lightning path. An example of an excavated 5 m (17 foot) fulgurite is shown in Fig. 12.7; and Fig. 12.8 contains a photograph of an excavated fulgurite of 1 m length that extends from the ground surface to a buried distribution power cable. The latter fulgurite was produced by a triggered lightning flash (see Section 13.2) in an experiment described by Barker and Short (1996a,b,c). There is considerable evidence from excavated fulgurites that single channels in sandy soil may carry the bulk of the lightning current when the lightning is attracted to underground conductors, as illustrated in Fig. 12.8.

When lightning current attaches to a buried cable covered in an insulating jacket (as in Fig. 12.8), the jacket can be punctured in a major or a minor way. In the latter case, the damage may not disrupt the operation of the cable. Nevertheless, a delayed failure can occur in days, weeks, or even years following the lightning event, as the result of such initial damage. A punctured jacket may allow water to leak into the cable which can cause corrosion of the internal metal, leading to eventual cable failure from a short circuit. Figure 12.9 shows a formerly buried, lightning-punctured PVC conduit and the power cable that resided within the conduit, from the same triggered-lightning experiment referenced above. Clearly, the buried PVC conduit did not provide protection from lightning, but such conduits do provide protection from misplaced shovels and miscreant rodents.

In triggered lightning tests (Barker and Short 1996a,b,c), lightning was attracted to an underground cable from as far away as 10 m on either side of the cable. Since first stroke currents in natural flashes are generally larger than triggered lightning stroke currents by a factor of 2 or 3, it is not unreasonable to expect that any natural flash within 10 to 20 m horizontal distance (or perhaps more) of a buried cable would have a high likelihood of attaching to the cable under the soil conditions which existed in the triggered lightning experiment, a relatively high soil resistivity of 4000 ohm-meters characteristic of much of the southeastern United States. A cable located in a good portion of Florida with a ground flash density of about $10 \, \mathrm{km}^{-2} \, \mathrm{yr}^{-1}$, with similar sandy soil conditions (yielding a 10 to 20 m attractive radius), might experience about 0.2 to 0.4 lightning attachments per kilometer of cable per year. This incidence is the same as the number of strikes to a 10 m high overhead distribution line as estimated by Eq. (12.1) and one-half the number from

Fig. 12.7   A Florida fulgurite of about 5 m length excavated by the University of Florida
Lightning Research Group.

Eq. (12.2), both assuming the overhead line is in an open field. However, the
incidence of strikes to overhead lines is dependent on their shielding (by trees,
buildings, and other tall objects). In the literature, a line shielding factor approaching 1 represents an overhead line imbedded in very tall trees so that the line cannot
be hit directly. A shielding factor of 0 represents a completely exposed line in open
terrain that would receive the maximum number of direct strikes. When lightning

Fig. 12.8   An excavated Florida fulgurite produced by triggered lightning to an underground coaxial power cable having an insulating jacket, and directly buried about 1 m underground. The fulgurite attached to the cable, damaging its insulating jacket. Photograph by V. A. Rakov.

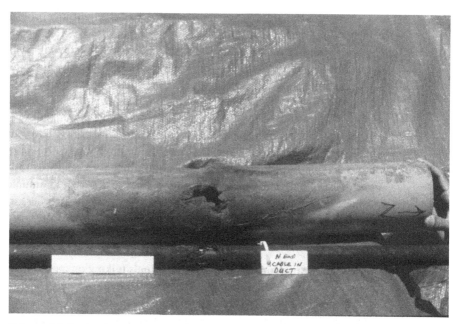

Fig. 12.9 Triggered lightning damage to an underground coaxial power cable and the PVC conduit inside which it was located. Both were buried at about 1 m depth. Note that, as expected, the power cable is damaged at the same location at which the lightning current punctured the hole in the PVC conduit.

strikes a tree or a structure, some or all of the lightning current can still potentially contact an underground cable. It follows that, considering the shielding that is usually present, the strike rates to overhead and underground lines should not be too dissimilar on average.

Airport runway lighting is energized by loops of buried power cables. Each individual light is powered by a small transformer whose primary is traversed by the cable loop current and whose secondary powers the individual lights. Failures from lightning are common in these underground lighting systems. The FAA-specified protective technique is a counterpoise buried about 0.15 m (6 inches) above the current loop. The counterpoise at that separation from the current loop may provide only limited protection in view of Eq. (12.3). The study of an airport lighting system in the presence of triggered lightning (see Chapter 13) is described by Bejleri *et al.* (2004).

12.4 Communication lines

Most telecommunication systems in common use, such as telephone and cable television, still transmit signals on metallic cables, but these cables are rapidly being replaced by wider-bandwith fiber optic lines. Fiber optic data transmission, above or under the ground, is immune to many of the lightning problems suffered by metal-wired systems, although fiber optic amplifiers and other signal processors may be subject to electronic damage as may accompanying metallic wires parallel-ing the fiber optic cables for support or strength. Through-the-air transmission, as in TV satellite systems, has no lightning problems, but the signals are blocked by the heavy precipitation associated with thunderstorms. In Section 12.3, we have considered the attractive radius of a buried power cable for a lightning strike to the Earth's surface. Those results are also applicable to buried telecommunication cables. In Section 12.2 we have discussed the voltages and currents on overhead conductors due to both nearby lightning and direct strikes. These results are also applicable to overhead telecommunication lines. Many telecommunication lines are shielded by overhead distribution power lines, as noted earlier, so that some aspects of the discussion of the shielding of overhead power lines by grounded wires found in Sections 12.1 and 12.2 are applicable to telecommunications lines shielded by power lines.

Injury and death from lightning of individuals talking on hard-wired phones are discussed in Section 7.3. Telephone signal wires in most countries are required by law to be protected from overvoltages due to both lightning effects and power line contact. As discussed in Section 7.3, this is generally accomplished with SPDs, most often gas tubes or carbon-block air-gap arresters that reduce the lightning-caused voltage to near zero via an electrical breakdown of the gas or air when the applied lightning voltage exceeds 500 to 1000 V (Section 6.3). These SPDs are placed between the signal wires and a grounding system (typically a ground rod) at the point that the wires enter a structure. Voltages entering the residential phone system above the 5 to 10 kV range will exceed the insulation level of most tele-phones and sparks may exit the phone handset. Five kilovolts can be caused by a current of 200 amperes traversing the SPD and entering a ground connection of 25 ohms, the maximum grounding resistance (that is seldom achieved in sandy soil) specified by the US National Electrical Code. Similar deleterious voltages can be caused by smaller currents encountering higher grounding resistances. Portable phones and cell phones do not have electric shock issues since they are not connected to outdoor wiring on which lightning-caused voltages can be induced.

According to Boyce (1977) the current induced in a single overhead telecommu-nication line by a large close lightning current may be hundreds of amperes with a pulse width of a few microseconds, decaying to tens of amperes with a pulse width of 10 microseconds or so after propagating over 1 kilometer of line. If flashover of the communication line insulators does not occur, SPDs on the line (if there are SPDs) will therefore discharge hundreds of amperes in the first tens of meters of line to tens of amperes a kilometer away. Insulation levels on telecommunication lines are, from a practical point of view, generally in the 50 kV range although design values are up to a factor of two higher. Boyce (1977) states that typical observed current waveshapes on long lines are 10/1000 μs while on short lines 2/100 μs is more representative, with multiple reflections occurring on long lines. Waveforms for testing the lightning immunity of communication lines are given in Section 6.2.

Unshielded overhead telecommunication lines will be struck at about the same rate as overhead power lines. Telecommunication lines that are suspended below power lines are fairly well protected from direct strikes and are also partially shielded from the induced effect of nearby strikes. Nevertheless, flashover from the power lines or the power line grounds to the telecommunication lines can occur. SPDs are essential to protect communication equipment on the line and at the line terminations.

Underground telecommunication cables may be directly buried or may be run in ducts or pipes or conduits, either metallic or more commonly plastic (PVC). The depth at which a telecommunication cable is buried has little effect on its suscept-ibility to lightning since within a meter or so of the surface, the practical depth at which cables can be buried, no immunity is obtained from direct strikes. The use of PVC pipes provides limited protection against lightning (see Fig. 12.9). As in the case of the underground power cables discussed in Section 12.3, parameters of importance to lightning effects on underground communication cables are the incidence level of lightning, the soil resistivity, the degree of shielding afforded by buildings, and the number and placement of other cables and metal pipes that can share the lightning current. Underground cable protection is provided by the shielding of metallic sheaths, ducts, and pipes, the use of SPDs, buried "shield" wires over the signal wires, and the dielectric strength of the signal conductor insulation.

Radio and television broadcast towers, police and fire emergency call facilities containing broadcast towers, microwave relay stations, and other tall towers associated with communication services are particularly exposed to lightning because of their height. Adequate protection for electronics and for individuals working in such facilities can only be obtained using the principles of topological shielding and transient protection discussed in Section 3.1. Grounding and bonding is seldom sufficient to limit deleterious voltages between separated but bonded grounds for the antenna tower and for the supporting structure containing com-munication electronics because of the inductive voltage drop in the bonding wires, which are necessarily exposed to a significant fraction of the total lightning current and typically extend over a distance of tens of meters (see example in Fig. 2.4). Thus the antenna tower voltage wave entering the electronics-housing structure on coaxial cables and waveguides may be very different from the voltages on consoles (metal tables at which equipment operators sit), the consoles being typically bonded to buried ring electrodes encircling the electronics-housing structure. Transient protection is essential in this situation in order to eliminate electrical shock to equipment operators, as well as to reduce damage to electronics. Seldom, however, is transient protection provided on enough of the voluminous amount of wiring in such facilities to solve the problem of voltage differences appearing between contiguous metal when the antenna tower is struck by lightning. Thus, individuals using hard-wired communication headsets and typing at metallic consoles in fire and police emergency call centers are particularly susceptible to serious shock.

References

Agrawal, A. K., Price, H. J. and Gurbaxani, S. J. 1980. Transient response of a field. *IEEE Trans. Electromagn. Compat.* **22**: 119–129.

Anderson, J. G. 1981. Monte-Carlo computer calculation of transmission line lightning performance. *IEEE Trans. Power Ap. Syst.* **80**: 414–420.

Barker, P. P. and Short, T. A. 1996a. Lightning effects studied: the underground cable program. *Transmission and Distribution World* (May), 24–33.

Barker, P. P. and Short, T. A. 1996b. Lightning measurements lead to an improved understanding of lightning problems on utility power systems. *Proc. 11 CESPSI*, Vol. 2, Kuala Lumpur, Malaysia, pp. 74–83.

Barker, P. and Short, T. 1996c. Findings of recent experiments involving natural and triggered lightning. *Panel Session Paper presented at 1996 Transmission and Distribution Conf.*, Los Angeles, California, September 16–20, 1996.

Bejleri, M., Rakov, V. A., Uman, M. A. *et al.* 2004. Triggered lightning testing of an airport runway lighting system. *IEEE Trans. Electromagn. Compat.*, **46**, No. 1, 96–101.

Borghetti, A., Nucci, C. A. and Paolore, M. 2007. An improved procedure for the assessment of overhead line indirect lightning performance and its comparison with IEEE Std. 1410 Method. *IEEE Trans. Power Delivery* **22**: 684–692.

Boyce, C. F. 1977. Protection of telecommunication systems. In *Lightning*, Vol. 2, *Lightning Protection*, ed. R. H. Golde. New York: Academic Press, pp. 793–829.

Currie, J. R., Choy, L. Ah. and Darveniza, M. 1971. Monte Carlo determination of the frequency of lightning strokes and shielding failures. *IEEE Trans. Power Ap. Syst.* **90**: 2305–2312.

Eriksson, A. J. 1987. The incidence of lightning strikes to power lines. *IEEE Trans. Power Delivery* **2**: 859–870.

IEEE 1996. *IEEE Standard 998: 1996*. IEEE guide for direct lightning stroke shielding of substations. New York: IEEE Power Engineering Society.

IEEE. 2004. *IEEE Standard 1410: 2004*. IEEE guide for improving the lightning performance of electric power overhead distribution lines. New York: IEEE Power Engineering Society.

IEEE. 2005. *IEEE Standard C62.11: 2005*. IEEE standard for metal-oxide surge arresters for AC power circuits (>1 kV). New York: IEEE Power Engineering Society.

Liew, A. C. and Darveniza, M. 1971. A sensitivity analysis of lightning performance for transmission lines. *IEEE Trans. Power Ap. Syst.* **90**: 1443–1451.

Liew, A. C. and Darveniza, M. 1982a. Calculation of the lightning performance of unshielded transmission lines. *IEEE Trans. Power Ap. Syst.* **101**: 1471–1477.

Liew, A. C. and Darveniza, M. 1982b. Lightning performance of unshielded transmission lines. *IEEE Trans. Pow. Appar. Syst.* **101**: 1478–1482.

McDermott, T. E. 2006. Line arrester energy discharge duties. *IEEE PES T&D Conference Proceedings*, May 21–24, pp. 450–454, doi:10.1109/TDC.2006.1668535.

Nucci, C. A. 1995. Lightning-induced voltages on overhead power lines, Part II: coupling models for the evaluation of the induced voltages. *Electra* **162**: 121–145.

Nucci, C. A. and Rachidi, F. 1995. On the contribution of the electromagnetic field components in field-to-transmission line interaction. *IEEE Trans. Electromagn. Compat.* **37**: 505–508.

Sargent, M. A. and Darveniza, M. 1967. The calculation of the double-circuit outage rate of transmission lines. *IEEE Trans. Power Ap. Syst.* **86**: 665–678.

Sargent, M. A. and Darveniza, M. 1970. Lightning performance of double circuit lines. *IEEE Trans. Power Ap. Syst.* **89**: 913–925

Schoene, J., Uman, M. A., Rakov, V. A. *et al.* 2007a. Direct lightning strikes to test power distribution lines. Part 1: Experiment and overall results. *IEEE Trans. Power Delivery*, **22**, No. 4 (October).

Schoene, J., Uman, M. A., Rakov, V. A. *et al.* 2007b. Direct lightning strikes to test power distribution lines. Part 2: Measured and modeled current division among multiple arresters and grounds. *IEEE Trans. Power Delivery*, **22**, No. 4 (October).

13 Lightning elimination

13.1 Modifying the cloud electrification process

In many circumstances it would be valuable to suppress or eliminate lightning. One potential way to do so would be to eliminate or reduce the source of lightning, the cloud charge. The primary charging mechanism in a thundercloud is thought to involve interactions between soft hail (graupel) falling under the influence of gravity and lighter ice crystals rising in the cloud's updrafts, all in the presence of liquid water drops not yet frozen but at an altitude where the temperature is colder than the freezing temperature ($32\,°F$ or $0\,°C$). Additional information is found in Section 1.1. Since cloud electrification is related to the falling precipitation within the cloud, and since this electrification is what produces lightning, it is reasonable to suppose that the amount of lightning is related to the amount of precipitation that is falling inside and below the cloud. Observationally, this is the case. In fact, one can often estimate the amount of rainfall (melted ice forms) that will reach the ground below a thunderstorm by counting the number of lightning flashes in the storm; the more lightning, the more rainfall (e.g., MacGorman and Rust 1998, Table 7.6; Takayabu 2006; Gungle and Krider 2006). For example, for nine isolated thunderstorms in Florida, Gungle and Krider (2006) found that each cloud-to-ground lightning flash was accompanied by an average rain volume of $2.6 \times 10^4 \pm 2.1 \times 10^4\,m^3$ (or a mean rainwater mass of $2.6 \times 10^7\,kg$) as measured by rain gauges on the ground. This result is consistent with previous measurements of summer thunderstorms made in Florida, Arizona, and New Mexico. A variety of other measurements, some made using radar to estimate rainfall, show similar or larger rain volumes (sometimes one to two orders of magnitude larger) per cloud-to-ground flash or per total cloud and cloud-to-ground flash. While our society would welcome a method to selectively eliminate lightning that might, for example, strike an explosive storage facility (see Fig. 2.5a,b) or injure spectators at a golf tournament, most farmers would not hesitate to contact their attorneys if they thought a lightning suppression system would also eliminate rain. There have been a number of experiments designed to *increase* the precipitation in selected clouds by releasing silver iodide or other materials into those clouds. In some cases, rain enhancement has been demonstrated. Such a demonstration can only be convincing if randomized experiments compare precipitation at the ground from many "seeded" clouds (seeded via aircraft releases of seeding material) with the precipitation from many "unseeded" clouds (via aircraft releases of inert material) with similar characteristics, where the experimenters do not know which clouds are being seeded and which are not. A study of this type generally requires five or more years in order to accumulate enough data to draw meaningful conclusions. Statistical methods must be used to evaluate whether there is an increase of rain at ground level from the seeded clouds since it is not known how the seeded clouds would have behaved had they not been seeded. A review of the present status of "weather modification" is given by List (2004) with references to the official positions taken on the subject by several major meteorological organizations. It is likely that experiments to modify a variety of aspects of the weather will be looked upon more favorably in the future in view of the potential changes in the climate associated with global warming. Mankind's continuing release of greenhouse gases into the Earth's atmosphere is an unintentional but measurable weather modification experiment that may well increase the global lightning activity (see Section 2.1).

Some experimental evidence exists that introducing metal needles into a thunderstorm can suppress cloud charging and hence lightning (Holitza and Kasemir 1974, Kasemir *et al.* 1976, Maddox *et al.* 1997), but sufficiently detailed experiments to test this hypothesis have not been undertaken.

13.2   Artificial initiation of lightning

Lightning can be artificially initiated from natural thunderstorms via action taken on the ground, resulting in a temporary decrease in cloud charge. One mechanism for such artificial initiation is the rocket-and-wire "triggering" of lightning illustrated in Fig. 13.1, Fig. 13.2, and Fig. 13.3. (The artificial initiation of lightning by in-flight aircraft and launch vehicles is discussed in Section 9.1 and illustrated in Fig. 9.2.) While artificially initiated ("triggered") lightning can drain thundercloud charge via lightning flashes that would not have occurred otherwise, the triggering of individual lightning flashes apparently does not significantly inhibit the ongoing process of cloud electrification. In fact, a typical thundercloud can regenerate the charge lost to a natural or triggered lightning in 10 to 20 s. Furthermore, the rocket-and-wire triggering process only works well during optimum conditions of relatively high electric fields at ground (typically $5 \, \text{kV m}^{-1}$ or higher, as measured with an electric field mill, see Section 8.1), and relatively low natural-lightning flash rate such that the high electric fields near ground remain relatively constant. In typical triggering campaigns, about 50 percent of rocket firings have successfully initiated lightning flashes. Even if the rocket-and-wire triggering process were effective as a cloud-charge drainage technique, it is probably too cumbersome an operation to be practical because there would typically be kilometers of unspooled, unexploded triggering wire draped over a rocket launching site from unsuccessful triggering attempts.

The mechanisms of rocket-and-wire triggering are shown in Fig. 13.1 in a sequence of six panels, increasing in time from left to right. Typically, when the

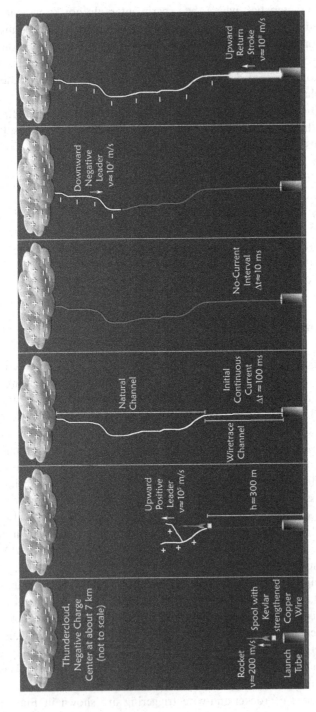

Fig. 13.1   Processes involved in the artificial initiation of lightning by rocket-and-wire triggering. Drawing by Jens Schoene.

Fig. 13.2   A photograph (time exposure) of rocket-and-wire triggered lightning. Shown is 200 to 300 m of luminous, straight wire with a tortuous lightning channel above. The triggering was done from ground level at the University of Florida's International Center for Lightning Research and Testing (ICLRT) at Camp Blanding, Florida.

rocket has lifted the trailing wire (in all recent studies the conducting wire is unspooled from the rocket) to a height of several hundred meters in a time of 2 or 3 s, electrical breakdown occurs at the top of the wire. For a several-hundred-meter length of vertical, grounded, conducting wire, the ambient electric field at and above the wire top from the negative cloud charge overhead is enhanced to a level sufficient to launch a positively charged leader upward from the wire top toward the negative cloud charge. When the positively charged upward leader from the rocket wire enters the negative cloud charge, it provokes an "initial continuous current" (ICC) of some hundred of amperes flowing for tenths of a second between the cloud charge and the ground. The triggering wire generally explodes or melts during the upward propagation of the positive leader prior to the ICC. When the ICC ceases to flow, the flash may end, or, more favorably for lightning research or testing, a dart leader may traverse the previous ICC channel in a downward direction from cloud charge to ground, followed by the propagation of a return stroke from the ground up the negatively charged leader channel (just as described in Section 1.3 for subsequent strokes in natural lightning). Triggered lightning strokes have been observed to be very similar, if not identical, to the subsequent strokes in natural lightning. The return stroke current from a triggered lightning flash is shown in Fig. 2.2.

Fig. 13.3   A photograph (time exposure) taken 30 m from lightning triggered to a 10 m tower at the ICLRT. The luminous straight channel on the left, above the rocket-launching tubes, is the remains of the bottom of the triggering wire. The wire luminosity is blown to the right by the wind. The 10 or so tortuous channels, also separated by the wind, have been traversed by 10 or so dart leader/return stroke sequences, each separated by some tens of milliseconds. The current associated with the triggered lightning, in the experiment shown, was directed by a hard-wired connection from the bottom of the triggering wire to either a test power line or a test residential structure in order to determine their lightning susceptibility.

Rocket-triggered lightning has many features in common with upward-initiated lightning from tall structures (Section 1.1, Fig. 1.2, and Fig. 1.4). In rocket-triggered lightning, the several hundred meters of vertical, grounded, conducting wire carried aloft by the rocket provides the same function as the tall, stationary, conducting structure. In both cases, an upward-moving leader from the grounded object is followed by an ICC, and these two processes replace the downward-moving stepped leader and upward-moving first return stroke of natural lightning. In both cases, there may or may not be dart leader/return stroke sequences that follow the ICC. These sequences are present in about half of both natural and triggered upward-initiated flashes. The fundamental difference between the two cases is that the initiating electric field for rocket triggering must be relatively steady whereas the initiating field for a stationary tall structure is thought to be a rapid electric field change caused by an overhead cloud discharge.

It has often been suggested that a laser beam should be the ideal substitute for the rocket wire in order to achieve more rapid and wire-free triggering. Other forms of beamed radiation such as microwaves have also been suggested. Apparently no laser to date has been able to trigger lightning, although there have been numerous attempts. This is probably because no laser system to date has been able to render hundreds of meters of air sufficiently conducting for a long enough period of time (as does the conducting wire used in rocket triggering) in order that ground potential be present at the top of the laser beam, leading to sufficient enhancement of the ambient electric field there to launch an upward leader. Further, most laser beams are absorbed or scattered by precipitation, precipitation being a relatively common feature of thunderstorms. Even if lasers could eventually be used to trigger lightning, it is still questionable whether such triggering would make a serious dent in Mother Nature's electrification operation.

Lightning was apparently artificially initiated by plumes of water thrown into the air from an underwater explosion in Chesapeake Bay. A frame from the movie showing that triggered lightning is given in Fig. 20.5 of Rakov and Uman (2003). It follows that lightning can likely be triggered by water jets similar to those produced by the fire boats that operate in some big-city harbors.

A detailed discussion of rocket-triggered lightning and other prospective types of artificially initiated lightning is found in Chapter 7 of Rakov and Uman (2003).

13.3   Can thunderstorms be steadily discharged?

Based on his laboratory experiments in discharging previously charged bodies through the air by approaching them with grounded, pointed conductors, Benjamin Franklin suggested in 1751 that it might be possible to discharge the cloud charge using grounded, pointed lightning rods placed on structures (Cohen 1990). Modern calculations indicate that the laboratory process can not be scaled up in this way because of the relatively small amount of charge released by pointed conductors, the relatively long time necessary for the air ions created near the pointed conductors to ascend the distance to the cloud charge, and the often relatively large horizontal motion of the air ions induced by wind (Uman and Rakov 2002). Nevertheless, there are a number of commercial "lightning elimination" systems in use today that advertise or imply such a mode of operation.

The primary claim of the proponents of lightning elimination systems (which more recently have been called "charge transfer systems") is that those systems produce conditions under which lightning either does not occur or cannot strike the protected structure, as opposed to the conventional approach of intercepting the imminent lightning strike and rendering it harmless by providing a non-destructive path for the lightning current to flow to ground. Lightning elimination systems include one or more elevated arrays of sharp points, often similar to barbed wire, that are installed on or near the structure to be protected. Examples are shown in Fig. 13.4 and Fig. 13.5. These arrays are connected to grounding electrodes via down conductors, in exactly the same manner as in conventional lightning protection systems. The principle of operation of lightning elimination systems (according to their proponents) is that the charge released via corona discharge at the sharp points will either (1) discharge the overhead thundercloud, thereby eliminating any

Fig. 13.4   A lightning elimination device claimed to protect a State of Florida truck weighing station on Interstate 75 in central Florida. Photograph by Derek Uman.

possibility of lightning (this is why such arrays are sometimes referred to as "dissipation arrays") or (2) discourage a downward-moving leader from attaching to the array and to the protected structure by reducing the electric field near the array and, hence, suppressing the initiation of upward-connecting leaders in the region near the array.

Fig. 13.5   Another of several types of "air terminals" intended to eliminate lightning from the area. Photograph by Derek Uman.

In addition to Benjamin Franklin's work on corona discharge, the idea of using multiple-point corona discharge to "silently" discharge thunderclouds and thus prevent lightning was also proposed in 1754 by Czech scientist Prokop Divisch (Müller-Hillebrand 1962) who constructed a "machina meteorologica" with over 200 sharp points installed on a 7.4-m-high wooden framework. A patent for a multiple-point lightning elimination system was issued in 1930 to J. M. Cage of Los Angeles, California (Hughes 1977). The patent describes the use of point-bearing wires suspended from a steel tower to protect petroleum storage tanks from lightning. A similar system, commonly referred to as a dissipation array system (DAS) or a charge transfer system (CTS), has been commercially available since 1971 although the product name and the name of the company that marketed it have changed over time (Carpenter 1977; Carpenter and Auer 1995). Most lightning elimination systems were originally designed for tall communication towers, but recently such systems have been applied to a wide range of systems and facilities including electrical substations, power lines, airports, and the Florida truck weighing station shown in Fig. 13.4.

Many researchers (e.g., Zeleny 1934, Golde 1977) have pointed out that natural corona from trees and grass can exceed the corona discharged from a dissipation array without inhibiting lightning. Further, as noted above, the corona charge from the dissipation arrays will be blown horizontally by the prevailing wind, away from both the structure to be protected or the cloud to be discharged, and the vertical motion of the meager number of charged particles resulting from the corona discharge is relatively slow on the scale of a charging thunderstorm. One umbrella dissipater, similar to that shown in Fig. 13.4, has been described by Bent and Llewellyn (1977) as being constructed of about 300 m of barbed wire wrapped spirally around the frame of a 6-m-diameter umbrella. The barbed wire has 2 cm barbs with four barbs separated by 90° placed every 7 cm along the wire. Mousa (1998) describes a ball dissipater, a barbed power line overhead shield wire, a conical barbed wire array, a cylindrical dissipater, a panel dissipater (fakir's bed of nails), and a doughnut dissipater. A "brush eliminator" is shown in Fig. 13.5. Dissipation array manufacturers list many reputable customers who report a cessation of lightning-caused damage after installation of the dissipation system, an installation that often includes improvement of grounding and addition of surge protection. In principle, lightning elimination systems can provide conventional lightning protection; that is, they can intercept a lightning strike and direct its current into the ground without damage to themselves or to the protected structure if there is sufficient coverage of the structure by arrays (air terminals). Further, damage to electronics within the structure can be eliminated or minimized by way of the installation of surge protective devices and good grounding, this protective effect having nothing to do with lightning elimination.

13.4   Can lightning be inhibited?

Golde (1977) has suggested that dissipation arrays installed on tall structures (typically communication towers) will inhibit upward lightning flashes (initiated by leaders that propagate upward from the tall structure into the cloud – see Section 13.2) by modifying the needlelike shape of the structure tops to a shape that has a less pronounced electric-field-enhancing effect. While this suggestion is not unreasonable, there are no measurements to support it. Upward lightning discharges occur from objects greater than 100 m or so in height (above flat terrain) and most lightning associated with objects of height above 400 to 500 m or so is upward (Section 1.1, Fig. 1.2, Fig. 1.4). In this view, dissipation arrays would inadvertently reduce the probability of occurrence of these upward flashes, which represent the majority of flashes to very tall towers. The upward flashes, as noted in Section 13.2, contain initial continuous current and often contain subsequent strokes similar to those in normal cloud-to-ground lightning. These strokes have the potential to damage electronics, damage that can be minimized by the use of surge protection devices (see Chapter 6). The view of Golde (1977) has been expanded upon by Mousa (1998), who argues that the suppression of upward flashes will be particularly effective for towers of 300 m height or more and that dissipation arrays will have no effect whatsoever on the frequency of strikes to smaller structures such as power substations and transmission line towers.

Observations exist of lightning strikes to dissipation arrays. In 1988 and 1989 the Federal Aviation Administration (FAA) conducted studies at three Florida airports of the performance of dissipation arrays relative to conventional lightning protection systems (FAA 1990). An umbrella dissipation array installed on the central tower of the Tampa International Airport was struck by lightning on August 27, 1989, as shown by video and current records (FAA 1990, appendix E). Carpenter and Auer (1995) have disputed the findings of FAA (1990), and Mousa (1998) has reviewed the attempts of the dissipation array manufacturer to suppress FAA (1990). Additional lightning strikes to dissipation arrays are described by Durrett (1977), Bent and Llewellyn (1977), and Rourke (1994). The former two references describe strikes to towers protected by dissipation arrays at the Kennedy Space Center, Florida, and at Eglin Air Force Base, Florida, respectively. Rourke (1994) considers lightning strikes to a nuclear power plant. The plant was struck by lightning three times in two years, 1988 and 1989, before having dissipation arrays installed. After dissipation array installation, the plant was also struck three times in two years, 1991 and 1992. Rourke (1994) notes that "there has been no evidence that lightning dissipation arrays can protect a structure by dissipating electric charge prior to the creation of the lightning."

Kuwabara et al. (1998) reported on a study of dissipation array systems that were installed in summer 1994 atop two communication towers on the roof of a building in Japan. Kuwabara et al. (1998) state that the dissipation array "was not installed per the manufacturer's recommendations as a result of the building construction conditions in Japan." Measurements of lightning current waveforms during strikes to the towers were made prior to the installation of dissipation arrays, from winter 1991 to winter 1994, and after the installation, from winter 1995 to winter 1996. Additionally, six direct strikes to the towers with the arrays installed were photographed between December 1997 and January 1998. Twenty-six lightning current waveforms were recorded in the three years before installation of the dissipation arrays and 16 in the year or so after installation. The statistical distribution of peak currents was essentially the same before and after installation. Estimated peak currents varied from 1 to 100 kA. Kuwabara et al. (1998) state that after installing the dissipation arrays, improving the grounding, and improving the surge protection in summer 1994, "malfunctions of the telecommunications system caused by lightning direct strike have not occurred," whereas they were common before. Apparently, the presence of the dissipation arrays neither prevented lightning strikes nor changed the characteristics of the lightning stroke current, while the equipment damage was eliminated by means of improved surge protection and grounding.

References

Bent, R. B. and Llewellyn, S. K. 1977. An investigation of the lightning elimination and strike reduction properties of dissipation arrays. *Review of Lightning Protection Technology for Tall Structures*, ed. J. Hughes. Publ. AD-A075 449. Arlington, VA: Office of Naval Research, pp. 149–241.

Carpenter, R. B. 1977. 170 system years of guaranteed lightning protection. *Review of Lightning Protection Technology for Tall Structures*, ed. J. Hughes Publ. AD-A075 449. Arlington, VA: Office of Naval Research, pp. 1–23.

Carpenter, R. B. and Auer, R. L. 1995. Lightning and surge protection of substations. *IEEE Trans. Ind. Appl.* **31**, 162–174.

Cohen, I. B. 1990. *Benjamin Franklin's Science*. Harvard: Harvard University Press.

Durrett, W. R. 1977. Dissipation arrays at Kennedy Space Center. *Review of Lightning Protection Technology for Tall Structures*, ed. J. Hughes. Publ. AD-A075 449. Arlington, VA: Office of Naval Research, pp. 24–52.

FAA 1990. 1989 Lightning protection multiple discharge systems tests: Orlands, Sarasota and Tampa, Florida. FAATCT16 Power Systems Program, ACN Final Report. Federal Aviation Administration.

Golde, R. H. 1977. The lightning conductor. In *Lightning Protection*, Vol. 2: *Lightning*, ed. R. H. Golde. London and New York: Academic Press, pp. 545–576.

Gungle, B. and Krider, E. P. 2006. Cloud-to-ground lightning and surface rainfall in warm-season Florida thunderstorms. *J. Geophys. Res.* **111**, doi:10.1029/2005JD006802.

Holitza, F. J. and Kasemir, H. W. 1974. Accelerated decay of thunderstorm electric fields by chaff seeding. *J. Geophys. Res.* **79**: 425–429.

Hughes, J. 1977. Introduction to *Review of Lightning Protection Technology for Tall Structures*, ed. J. Hughes. Publ. AD-A075 449. Arlington, VA: Office of Naval Research, pp. i–iv.

Kasemir, H. W., Holitza, F. J., Cobb, W. E. and Rust, W. D. 1976. Lightning suppression by chaff seeding at the base of thunderstorm. *J. Geophys. Res.* **81**: 1965–1970.

Kuwabara, N., Tominaga, T., Kanazawa, M. and Kuramoto, S. 1998. Probability occurrence of estimated lightning surge current at lightning rod before and after installing dissipation array system (DAS). *IEEE Electromag. Compat. Int. Symp. Record*, Denver, CO., pp. 1072–1077.

List, R. 2004. Weather modification – a scenario for the future. *Bull. Am. Meteorol. Soc.* **85**: 51–63, doi:10.1175/BAMS-85-1-51.

MacGorman, D. R. and Rust, W. D. 1998. *The Electrical Nature of Storms*. New York: Oxford University Press.

Maddox, R. A., Howard, K. W. and Dempsey, C. L. 1997. Intense convective storms with little or no lightning over central Arizona: a case of inadvertent weather modification? *J. Appl. Meteorol.* **36**: 302–314.

Mousa, A. M. 1998. The applicability of lightning elimination devices to substations and power lines. *IEEE Trans. Power Delivery* **13**: 1120–1127.

Müller-Hillebrand, D. 1962. The protection of houses by lightning conductors – an historical review. *J. Franklin Inst.* **273**: 35–44.

Rakov, V. A. and Uman, M. A. 2003. *Lightning: Physics and Effects*. Cambridge: Cambridge University Press.

Rourke, C. 1994. A review of lightning-related operating events at nuclear power plants. *IEEE Trans. Energy Conver.* **9**: 636–641.

Takayabu, Y. N. 2006. Rain-yield per flash calculated from TRMM PR and LIS data and its relationship to the contribution of tall convective rain. *Geophys. Res. Lett.* **33**: L18705, doi:10.1029/2006GL027531.

Uman, M. A. and Rakov, V. A. 2002. A critical review of non-conventional approaches to lightning protection. *Bull. Am. Meteorol. Soc.* **83**: 1809–1820.

Zeleny, J. 1934. Do lightning rods prevent lightning? *Science* **79**: 269–271.

14 So, what do we know and what don't we know about lightning protection?

14.1 Does it work?

The most important thing we know about lightning protection is that if implemented in a manner consistent with existing standards, the protection will be successful most of the time. Many individual examples illustrate the value of lightning protection systems. Some are given below.

According to Schonland (1950):

The record of damage to churches, whose elevated steeples attract the lightning flash, is voluminous. . . . Perhaps the most famous of these structures is the Campanile of St. Mark in Venice which has had a very bad lightning history. It stands over 340 feet high in an area which, as already mentioned, experiences many thunderstorms. It was severely damaged by a stroke in 1388, at which time it was a wooden structure. In 1417 it was set on fire by lightning and destroyed. In 1489 it was again reduced to ashes. In 1548, 1565, and 1653 it was damaged more or less severely, and in 1745 a stroke of lightning practically ruined the whole tower. Repairs cost 8,000 ducats (3,000 pounds sterling in those days), but in 1761 and 1762 it was again severely damaged. In 1776 a Franklin rod was installed on it and no further trouble from lightning has occurred since.

We noted in Section 10.1 that William Snow Harris compiled a record of lightning damage to the unprotected wooden ships of the British Royal Navy. The lightning protection system for ships devised by Harris was adopted for trial purposes, beginning in 1830, on 11 Royal Navy vessels ranging in size from the 120-gun ship-of-the-line *Caledonia* to the 10-gun brig *Beagle*, on which Charles Darwin set sail in 1831. Those trials were successful, and the system proposed by Harris was formally adopted by the British Admiralty in 1841, with all Royal Navy vessels protected by the system by 1850. Serious lightning damage was virtually eliminated. See Section 10.1 for more details and references.

Kellogg (1912) provided overall statistics on lightning-caused damage to structures from the Iowa Farmers Mutual Fire Insurance Associations for the period from 1908 to 1911. Detailed data are provided for 1908 and 1909. He estimated that about 25 percent of significant rural buildings in Iowa had lightning protection at that time. In 1908 and 1909, buildings in Iowa without lightning protection systems suffered $81 077 in damage whereas buildings with such systems suffered only $1078 in damage. From 1908 to 1911, the ratio of the cost of damage in unprotected versus protected buildings was about 60:1. Kellogg states that in most cases of damage to buildings with rods "the cause was found to be defective or incomplete rodding." Taking into account that only about 25 percent of the structures had lightning protection, Kellogg calculated that the ratio of the cost of lightning damage to unprotected and protected structures was about 15:1.

Covert (1930) presented data on lightning fires in Iowa from 1919 to 1924. Almost 80 percent of the losses for unprotected structures were to barns, grain elevators, and other farm structures. He states that about half the structures in rural Iowa had lightning protection during the time considered. Of about $2 million in lightning-caused fire losses over the six years, 93 percent was to structures without lightning protection.

McEachron and Patrick (1940) reported on a study of lightning protection systems that began in the 1920s in Ontario, Canada.

A 10-year survey in the Province of Ontario, in Canada, disclosed that during the period covered, 10,079 lightning fires took place in structures not equipped with lightning rods, while only 60 such fires occurred in buildings with lightning rod systems of protection. Of these 60 fires, it was found that many were started in structures equipped with improper lightning rods, or rods in bad condition because of poor maintenance. It is safe to say today that a lightning rod system practically eliminates the chance of damage from a stroke, although it will not prevent the stroke itself . . .

According to Viemeister (1972):

In 1923 the National Board of Fire Underwriters inaugurated a system for monitoring the installation of lightning protection systems through Underwriter's Laboratories. A Master Label is granted to a system that meets a stringent set of requirements. Since the state of the Master Label program, more than 240,000 labels have been awarded, and less than one-tenth of 1 per cent have been reported damaged by lightning. Investigators found that in the majority of damage cases the protection system was either in poor condition or the building had been updated without appropriate updating of the system.

According to McEachron (1952):

A survey by [Office of the Chief of] Ordinance [US Army] for the period from 1944 through 1948 shows the following: a. Protected structures were struck 330 times; damage negligible. b. Unprotected structures were struck 52 times; damage exceeded $130,000.

14.2   How well does it work?

Given that lightning protection systems specified by the standards do work (albeit not perfectly), exactly how well do they work and how can their performance be optimized? Will it ever be possible, for example, to compare experimentally or theoretically the relative performance of an air terminal composed of a given conducting mesh laid directly on a roof (approved by the international IEC standard but not by the US NFPA standard) with the performance of a system of electrically connected rods of a particular separation? Is the theory described in Sections 3.3, 3.4, and 4.2 sufficient for proper lightning protection design or is the lightning attachment process just too complex to be amenable to such simple modeling? Can the theory in Sections 3.3, 3.4 and 4.2 be used to describe the failure rate of a protection system with any degree of confidence? Certainly an examination of Fig. 1.8, a streak photograph of the attachment process for one case (there are only a few such cases published), would lead to considerable insecurity in even defining a striking distance that is uniquely related to the following return stroke current. In fact, the earlier step in the theoretical development, the relationship between the striking distance and the critical electric field present between the leader and the object to be struck, is itself a vast over-simplification as we have discussed in Section 3.3. While a relationship between the striking distance and the return stroke peak current is physically reasonable, at least on average, any such relationship clearly rests on a very shaky theoretical foundation. Evidence for this, if any is needed, is that the factors A and b found in the literature for Eq. (14.1), the most common expression relating the striking distance d to the peak current I_p,

$$d = A I_p^b \tag{14.1}$$

where d is in meters and I_p is in kiloamperes, exhibit a considerable range (Section 3.3 and Fig. 3.6), and there is not much unambiguous experimental evidence for the validity of the "best" A and b, if there be such.

If, however, Eq. (14.1) is a reasonable mathematical representation of the attachment processes on the average, the choice of A and b primarily serves to determine the fraction of lightning flashes for which the protection system may not function properly; that is, to determine what percentage of the fairly well-known distribution of first return stroke peak currents are smaller than the peak current associated with the striking distance (rolling sphere radius) specified by the chosen values of A and b (see Section 4.2). But even these smaller currents can only strike the structure in limited areas when it is completely protected against larger currents. It follows that the percentage of smaller currents for which failure is possible is larger than the percentage for which failure will actually occur; even the smallest lightning current can strike an air terminal if the stepped leader is in the proper location. Further, when lightning rods are used as air terminals, they are electrically connected with bonding cables, and these cables effectively form a mesh air termination system that is present in addition to the rods, thus decreasing the probability of overall system failure.

Is it perhaps telling that standards for lightning protection of small structures have changed relatively little from the time that the existence of the stepped leader and attachment process were unknown to the time following the development of the various versions of the electrogeometric model? Certainly, the electrogeometric model has illuminated our understanding of how and why lightning strikes, particularly how it strikes relatively tall structures (see Fig. 3.9a,b). Have the recent refinement and standardization of the rules for lightning protection using theoretical models merely provided justification for the protection that was developed via experience prior to our acquiring detailed knowledge of the lightning processes? Is there any reason to use the more involved rolling sphere method rather than the simpler cone of protection method when protecting small structures, those of height less than 10 to 20 meters (see Sections 3.4 and 4.2)?

Clearly, a better understanding of the attachment process is critical to advancing the theory of lightning protection, and hence answering at least some of the question posed in this section. This better understanding can only come through long-term research. Even with the most modern of research equipment to acquire data on the attachment process, that research will necessarily take many years since the lightning must strike in the volume of space being monitored. While many such volumes can be simultaneously monitored, the researcher is still operating on the schedule of the lightning in being able to accumulate enough events, each different, to characterize sufficiently the average attachment process and its range of variation. If near-certain lightning protection is needed at present, topological shielding with transient protection (see Section 3.1) is the only relatively foolproof approach available. Such an approach is best adopted in the design stage of the structure since retrofitting such protection is often impractical, expensive, or both.

References

Covert, R. N. 1930. Protection of buildings and farm property from lightning. *U.S. Department of Agriculture Farmers Bulletin No. 1512.* Issued Nov. 1926, revised Aug. 1930. Washington, DC.

Kellogg, E. W. 1912. The use of metal conductors to protect buildings from lightning. *University of Missouri Bulletin No. 7, Engineering Experiment Station*, Vol. 3, No. 1. Columbia, Missouri.

McEachron, K. B. 1952. Lightning protection since Franklin's day. *J. Franklin Inst.* 235: 441–470.

McEachron, K. B. and Patrick, K. G. 1940. *Playing with Lightning* New York: Random House.

Schonland, B. F. J. 1950. *Flight of Thunderbolts.* Oxford Clarendon Press.

Viemeister, P. E. 1972. *The Lightning Book.* Cambridge, MA: MIT Press.

Index